Data Augmentation with Python

Enhance deep learning accuracy with data augmentation
methods for image, text, audio, and tabular data

Duc Haba

BIRMINGHAM—MUMBAI

Data Augmentation with Python

Group Product Manager: Ali Abidi

Publishing Product Manager: Dinesh Chaudhary

Senior Editor: Sushma Reddy

Technical Editor: Devanshi Ayare

Copy Editor: Safis Editing

Project Manager: Kirti Pisat

Project Coordinator: Farheen Fathima

Proofreader: Safis Editing

Indexer: Hemangini Bari

Production Designer: Joshua Misquitta

Marketing Coordinator: Shifa Ansari & Vinishka Kalra

First published: April 2023

Production reference: 1270423

Published by Packt Publishing Ltd.

Livery Place

35 Livery Street

Birmingham

B3 2PB, UK.

ISBN 978-1-80324-645-1

www.packtpub.com

They say love is immeasurable, but I say love is measured by two eggs sunny side up, ham, toast, and a cup of tea. My dad has made this breakfast meal for me in the morning, afternoon, evening, and late nights. From my college days till today, he made them with love. Thanks, Dad. :-)

– Duc Haba

Foreword

I recently had the distinct pleasure of interviewing Duc (pronounced "Duke") regarding his lifelong passion for Artificial Intelligence, as part of an effort to promote an AI Hackathon he was leading. I'd known Duc for several years at this point — we're both engineering leaders at a premier agency in Silicon Valley. But I had no idea that he was an early pioneer in the AI industry. I was often surprised — even astonished — by the depth and breadth of his experience as we spoke about the history and future of AI. Equally impressive were the caliber of friends he's made along the way, including global AI leaders.

What impressed me most about Duc, however, was the man himself. Yes, he's in a rarified stratum of talent and capability, and everyone who knows him is aware of this. But despite his prodigious talent and world-class AI pedigree, Duc is a warm-hearted person, entirely down-to-earth, approachable, and even charming. He treats everyone as a peer, extending a cheerful hand of friendship. This great warmth of character permeates everything he does.

Not only is he blazing trails into unconquered technical territories, but he's also carving those trails wide and clean so that others can follow with ease. He holds two sacred goals in mind. The first is charting a positive course into the potential of AI to unlock world-changing solutions. The second is empowering as many people as possible to join him on his journey, both to share in the wonder of it all and to provide together a fabric of conscience that wraps around AI as a living safeguard against abuse.

I believe you'll find Duc's sacred goals, coupled with his peculiar strengths, on full display in this book. First, he tackles the signature pain point of AI, which is a dearth of data. You see, the ultimate accuracy of every AI model depends entirely upon the quality and quantity of data that the model is derived from, but data can be prohibitively expensive, or impossible to gather at scale. Duc solves this dearth of data for every major data type: image, text, audio, and tabular. This is the first of Duc's goals, to unlock the peerless power of AI.

But Duc doesn't stop there. He meticulously charts and documents his techniques to make them readily available to you, dear reader, to share his most important discoveries. He makes it easy to adopt his groundbreaking work because that's the second of his goals, to democratize AI for the betterment of us all.

The techniques in this book will exponentially expand your data sets and thus drastically improve the accuracy of your AI models. They're a ready-made bridge to your own AI dreams.

If you look ahead, just up that wide, clean path into the magical world of AI, you can see Duc standing there with a big warm smile, waving you on.

Jonmar

Engineering and Outreach at YML, Founder of Varlio, TEDx speaker, Featured by Apple, Rolling Stone, and Guitar World magazine.

Contributors

About the author

Duc Haba is a lifelong technologist and researcher. He has been a programmer, Enterprise Mobility Solution Architect, AI Solution Architect, Principal, VP, CTO, and CEO. The companies range from startups and IPOs to enterprise companies.

Duc's career started with Xerox Parc, researching and building expert systems (ruled-based) for copier diagnostic, and skunk works for the USA DOD. Afterward, he joined Oracle, following Viant Consulting as a founding member. He dove deep into the entrepreneurial culture in Silicon Valley. There were slightly more failures than successes, but the highlights are Viant and RRKidz. Currently, he is happy working at YML.co as the AI Solution Architect.

The book is only possible with the support of my family, fellow researchers, and a small gang of professionals at Packt Publishing. Still, above all else, I hope you enjoy reading the book and hacking the Python Notebook as much as I enjoyed writing it.

About the reviewers

Krishnan Raghavan is an IT Professional with over 20+ years of experience in software development and delivery excellence across multiple domains and technology ranging from C++ to Java, Python, Data Warehousing, and Big Data Tools and Technologies.

When not working, Krishnan likes to spend time with his wife and daughter besides reading fiction, nonfiction and technical books. Krishnan tries to give back to the community by being part of GDG – Pune Volunteer Group helping the team oraganize events. Currently, he is unsuccessfully trying to learn how to play the guitar. :)

You can connect with Krishnan at mailtokrishnan@gmail.com or via LinkedIn: www.linkedin.com/in/krishnan-raghavan.

I would like to thank my wife Anita and daughter Ananya for giving me the time and space to review this book.

Rajvardhan Oak is an Applied Scientist at Microsoft and a Ph.D. candidate in Computer Science at UC Davis, advised by the esteemed Professor Zubair Shafiq. Rajvardhan graduated from UC Berkeley with a Masters in Information Management and Systems, where he gained valuable experience in applying ML to security issues, such as detecting fake news, hate speech, adversarial machine learning, and phishing and spam detection. With an impressive resume, Rajvardhan has worked with industry giants such as Facebook and IBM. He has also been involved in Sec-ML research at UC Berkeley, IIT Kharagpur, and Princeton University.

Vitor Bianchi Lanzetta (@vitorlanzetta) has a master's degree in Applied Economics (University of So PauloUSP) and works as a data scientist in a tech start-up named RedFox Digital Solutions. He has also authored a book called R Data Visualization Recipes. The things he enjoys the most are statistics, economics, and sports of all kinds (electronics included). His blog, made in partnership with Ricardo Anjoleto Farias (@R_A_Farias), can be found at ArcadeData dot org, they kindly call it R-Cade Data.

Bhavan Jasani works as an Applied Scientist at Amazon Web Services AI in San Francisco. His work focuses on multi-modal learning and vision-language. Before that, he did his Master's in Robotics by Research from Robotics Institute, Carnegie Mellon University, Pittsburgh working on multi-modal emotion recognition and visual question answering. He was also a research staff at Nanyang Technological University, Singapore, working on embedded computer vision. He has reviewed and published his work in leading computer vision conferences, including ICCV and ECCV, as well as IEEE journals.

Table of Contents

Part 2: Image Augmentation

3

Image Augmentation for Classification   61

4

Image Augmentation for Segmentation   123

Part 3: Text Augmentation

5

6

Part 4: Audio Data Augmentation

7

8

Audio Data Augmentation with Spectrogram   291

Part 5: Tabular Data Augmentation

9

Tabular Data Augmentation   323

Preface

Unleash the power of data augmentation for AI and Generative AI by utilizing real-world datasets. Improve your model's accuracy and expand images, texts, audio, and tabular using over 150 fully functional object-oriented methods and open-source libraries.

Who this book is for

This book is intended for individuals interested in the AI discipline, including data scientists and students. While advanced AI or deep learning skills are not required, familiarity with Python programming and Jupyter Notebooks is necessary.

What this book covers

Chapter 1, Data Augmentation Made Easy, is an introduction to data augmentation. Readers will learn the definition of data augmentation, data types, and its benefits. Furthermore, the readers will learn how to select the appropriate online Jupyter Python Notebook or install it locally. Finally, Chapter 1 concludes with a discussion on coding conventions, GitHub access, and the foundation of Object-Oriented class code, named Pluto.

Chapter 2, Biases in Data Augmentation, defines the computation, human, and systemic biases with plenty of real-world examples to illustrate the differences between these types of biases. Readers will have the opportunity to practice identifying data biases by downloading three real-world image datasets and two text datasets from the Kaggle website to reinforce their learning. Once downloaded, readers will learn how to display image and text batches and discuss potential biases in the data.

Chapter 3, Image Augmentation for Classification, has two parts. First, readers will learn the concepts and techniques of augmentation for Image classification, followed by hands-on Python coding and a detailed explanation of the image augmentation methods with a safe level of image distortion. By the end of this chapter, readers will learn the concepts and hands-on techniques in Python coding for classification image augmentation using six real-world image datasets. In addition, they will examine several Python open-source libraries for image augmentation and write Python wrapper functions using the chosen libraries.

Chapter 4, Image Augmentation for Segmentation, highlights that both Image Segmentation and Image Classification are critical components of the Computer Vision domain. Image Segmentation involves grouping parts of an image that belong to the same object, also known as pixel-level classification. Unlike Image Classification, which identifies and predicts the subject or label of a photo, Image

Segmentation determines if a pixel belongs to a list of objects or tags. The image augmentation methods for segmentation or classification are the same, except segmentation comes with an additional mask or ground-truth image. Chapter 4 aims to provide continuing Geometric and Photometric transformations for Image Segmentation.

Chapter 5, Text Augmentation, explores text augmentation, a technique used in natural language processing (NLP) to generate additional data by modifying or creating new text from existing text data. Text augmentation can involve techniques such as character swapping, noise injection, synonym replacement, word deletion, word insertion, and word swapping. Image and Text augmentation has the same goal. They strive to increase the training dataset's size and improve AI prediction accuracy. In Chapter 5, you will learn about Text augmentation and how to code the methods in the Python Notebooks.

Chapter 6, Text Augmentation with Machine Learning, discusses an advanced technique that aims to improve ML model accuracy. Interestingly, text augmentation uses a pre-trained ML model to create additional training NLP data, creating a circular process. Although ML coding is beyond the scope of this book, understanding the difference between using libraries and ML for text augmentation can be beneficial. Chapter 6 will cover text augmentation with Machine Learning.

Chapter 7, Audio Data Augmentation, explains that similar to image and text augmentation, the objective of audio augmentation is to extend the dataset for gaining higher accuracy forecast or prediction in a Generative AI system. Audio augmentation is cost-effective and a viable option when acquiring additional audio files is expensive or time-consuming. Writing about audio augmentation methods poses unique challenges. The first is that audio is not visual like images or text. If the format is audiobooks, web pages, or mobile apps, we play the sound, but the medium is paper. Thus, we will transform the audio signal into a visual representation. Chapter 6 will cover Audio augmentation using Waveform transformation. You can play the audio file on the Python Notebook.

Chapter 8, Audio Data Augmentation with Spectogram, builds on the previous chapter's topic of audio augmentation by exploring additional visualization methods beyond the Waveform graph. An audio spectrogram is another visualizing method to see the audio components. The inputs to the spectrogram are a one-dimensional array of amplitude values and the sampling rate. They are the same inputs as the Waveform graph. An audio spectrogram is sometimes called sonographs, sonagrams, voiceprints, or voicegrams. The typical usage is for music, human speech, and sonar. A short standard definition is a spectrum of frequency maps with time duration. In other words, the Y-axis is the frequency in Hz or kHz, and the X-axis is the time duration in seconds or milliseconds. Chapter 8 will cover the audio spectrogram standard format, variation of a spectrogram, Mel-spectrogram, Chroma Short-time Fourier transformation (STFT), and augmentation techniques.

Chapter 9, Tabular Data Augmentation, involves taking data from a database, spreadsheet, or table format and extending it for the AI training cycle. The goal is to increase the accuracy of prediction or forecast, which is the same for image, text, and audio augmentations. Tabular augmentation is a relativelynew field for Data scientists. It is contrary to using analytics for reporting, summarizing, or forecasting. In analytics, altering or adding data to skew the results to a preconceived desired outcome is unethical. In data augmentation, the purpose is to derive new data from an existing dataset. The two goals are incongruent, but they are not. There will be a slight departure from the image, text, and audio augmentation format. We will spend more time in Python code studying the real-world tabular dataset.

To get the most out of this book

I designed this book to be a hands-on journey. It will be more effective to read a chapter, run the code on the Python Notebook, re-read the chapter's part that confused you, and jump back to hacking the code until the concept or technique is firmly understood.

Software/hardware covered in the book	Operating system requirements
Python	Chrome, Edge, Safari, or FireFox browser on Windows, macOS, or Linux.
Jupyter Notebook (Python Notebook)	
Python standard libraries, Panda, Matplotlib, and Numpy	
Python image, text, audio, and tabular data augmentation libraries.	

The default online Jupyter Notebook is the Google Colab. You need a Google account. For other online Jupyter Notebook, like Kaggle, Microsoft, or other online Jupyter Notebook, you need sign up or have an account to their services.

If you are using the digital version of this book, we advise you to type the code yourself or access the code from the book's GitHub repository (a link is available in the next section). Doing so will help you avoid any potential errors related to the copying and pasting of code.

Downloading real-world dataset from the Kaggle website requires a Kaggle username and key.

Download the example code files

You can download the example code files for this book from GitHub at `https://github.com/PacktPublishing/Data-Augmentation-with-Python`. If there's an update to the code, it will be updated in the GitHub repository.

We also have other code bundles from our rich catalog of books and videos available at `https://github.com/PacktPublishing/`. Check them out!

Download the color images

We also provide a PDF file that has color images of the screenshots and diagrams used in this book. You can download it here: `https://packt.link/FhpHV`

Conventions used

There are a number of text conventions used throughout this book.

`Code in text`: Indicates code words in text, database table names, folder names, filenames, file extensions, pathnames, dummy URLs, user input, and Twitter handles. Here is an example: "Using the `fetch_kaggle_comp_data()`, `fetch_df()`, and `draw_batch()` wrapper functions."

A block of code is set as follows:

```
pluto.remember_kaggle_access_keys("your_username_here",
  "your_key_here")
```

When we wish to draw your attention to a particular part of a code block, the relevant lines or items are set in bold:

```
# Instantiate Pluto
pluto = PackTDataAug("Pluto")
```

Any command-line input or output is written as follows:

```
!git clone https://github.com/duchaba/Data-Augmentation-with-Python
```

Bold: Indicates a new term, an important word, or words that you see onscreen. For instance, words in menus or dialog boxes appear in **bold**. Here is an example: "Next, go to the **Account** page, scroll down to the **API** section, and click on the **Create New API Token** button to generate the **Kaggle key**."

> **Tips or important notes**
> Appear like this.

Get in touch

Feedback from our readers is always welcome.

General feedback: If you have questions about any aspect of this book, email us at customercare@ packtpub.com and mention the book title in the subject of your message.

Errata: Although we have taken every care to ensure the accuracy of our content, mistakes do happen. If you have found a mistake in this book, we would be grateful if you would report this to us. Please visit www.packtpub.com/support/errata and fill in the form.

Piracy: If you come across any illegal copies of our works in any form on the internet, we would be grateful if you would provide us with the location address or website name. Please contact us at copyright@packt.com with a link to the material.

If you are interested in becoming an author: If there is a topic that you have expertise in and you are interested in either writing or contributing to a book, please visit authors.packtpub.com..

Share your thoughts

Once you've read *Data Augmentation with Python*, we'd love to hear your thoughts! Scan the QR code below to go straight to the Amazon review page for this book and share your feedback.

https://packt.link/r/1-803-24645-6

Your review is important to us and the tech community and will help us make sure we're delivering excellent quality content.

Download a free PDF copy of this book

Thanks for purchasing this book!

Do you like to read on the go but are unable to carry your print books everywhere?

Is your eBook purchase not compatible with the device of your choice?

Don't worry, now with every Packt book you get a DRM-free PDF version of that book at no cost.

Read anywhere, any place, on any device. Search, copy, and paste code from your favorite technical books directly into your application.

The perks don't stop there, you can get exclusive access to discounts, newsletters, and great free content in your inbox daily

Follow these simple steps to get the benefits:

1. Scan the QR code or visit the link below

https://packt.link/free-ebook/9781803246451

1. Submit your proof of purchase
2. That's it! We'll send your free PDF and other benefits to your email directly

Part 1: Data Augmentation

This part includes the following chapters:

1

Data Augmentation Made Easy

Data augmentation is essential for developing a successful **deep learning** (DL) project. However, data scientists and developers often overlook this crucial step. It is no secret that you will spend the majority of your project time gathering, cleaning, and augmenting the dataset in a real-world DL project. Thus, learning how to expand the dataset without purchasing new data is essential. This book covers standard and advanced techniques for extending image, text, audio, and tabular datasets. Furthermore, you will learn about data biases and learn how to code on **Jupyter Python Notebooks**.

Chapter 1 will introduce various data augmentation concepts, set up the coding environment, and create the foundation class. Later chapters will explain various techniques in detail, including Python coding. The effective use of data augmentation has proven to be the deciding factor between success and failure in **machine learning** (ML). Many real-world ML projects stay in the conceptual phase because of insufficient data for training the ML model. Data augmentation is a cost-effective technique that can increase the size of the dataset, lower the training error rate, and produce a more accurate prediction and forecast.

> **Fun fact**
>
> The car gasoline analogy is helpful for students who first learn about data augmentation and **artificial intelligence** (AI). You can think of data for the AI engine as the gasoline and data augmentation as the additive, such as the Chevron Techron fuel cleaner, that makes your car engine run faster, smoother, and further without extra petrol.

In this chapter, we'll define the data augmentation role and the limitations of extending data without changing its integrity. We'll briefly discuss the different types of input data, such as image, text, audio, and tabular data, and the challenges in supplementing it. Finally, we'll set up the system requirements and the programming style in the accompanying Python notebook.

I designed this book to be a hands-on journey. It will be most effective to read a chapter, run the code, re-read the part of the chapter that confused you, and jump back to hacking the code until you firmly understand the concept or technique that was presented.

You are encouraged to change or add new code to the Python notebook. The primary purpose of this book is interactive learning. So, if something goes wrong, download a fresh copy from the book's GitHub. The surest method to learn is to make mistakes and create something new.

Data augmentation is an iterative process. There is no fixed recipe. In other words, depending on the dataset, you select augmented functions and jiggle the parameters. A subject domain expert may provide insight into how much distortion is acceptable. By the end of this chapter, you will know the general rules for data augmentation, what type of input data can be augmented, the programming style, and how to set up a Python Notebook online or offline.

In particular, this chapter covers the following primary topics:

- Data augmentation role
- Data input types
- Python Notebook
- Programming styles

Let's start with the data augmentation role.

Data augmentation role

Data is paramount in any AI project. This is especially true when using the **artificial neural network** (**ANN**) algorithm, also known as **DL**. The success or failure of a DL project is primarily due to the input data quality.

One primary reason for the significance of data augmentation is that it is relatively too easy to develop an AI for prediction and forecasting, and those models require robust data input. With the remarkable advancement in developing, training, and deploying a DL project, such as using the **FastAI** framework, you can create a world-class DL model in a handful of Python code lines. Thus, expanding the dataset is an effective option to improve the DL model's accuracy over your competitor.

The traditional method of acquiring additional data is difficult, expensive, and impractical. Sometimes, the only available option is to use data augmentation techniques to extend the dataset.

> **Fun fact**
>
> Data augmentation methods can increase the data's size tenfold. For example, it is relatively challenging to acquire additional skin cancer images. Thus, using a random combination of image transformations, such as vertical flip, horizontal flip, rotating, and skewing, is a practical technique that can expand the skin cancer photo data.

Without data augmentation, sourcing new skin cancer photos and labeling them is expensive and time-consuming. The **International Skin Imaging Collaboration** (**ISIC**) is the authoritative data source for skin diseases, where a team of dermatologists verified and classified the images. ISIC made the datasets available to the public to download for free. If you can't find a particular dataset from ISIC, it is difficult to find other means, as accessing hospital or university labs to acquire skin disease images is laced with legal and logistic blockers. After obtaining the photos, hiring a team of dermatologists to classify the pictures to correct diseases would be costly.

Another example of the impracticality of attaining additional images instead of augmentation is when you download photos from social media or online search engines. Social media is a rich source of image, text, audio, and video data. Search engines, such as **Google** or **Bing**, make it relatively easy to download additional data for a project, but copyrights and legal usage are a quagmire. Most images, texts, audio, and videos on social media, such as **YouTube**, **Facebook**, **TikTok**, and **Twitter**, are not clearly labeled as copyrights or public domain material.

Furthermore, social media promotes popular content, not unfavorable or obscure material. For example, let's say you want to add more images of parrots to your parrot classification AI system. Online searches will return a lot of blue-and-yellow macaws, red-and-green macaws, or sulfur-crested cockatoos, but not as many Galah, Kea, or the mythical Norwegian-blue parrot – a fake parrot from the Monty Python comedy skit.

Insufficient data for AI training is exacerbated for text, audio, and tabular data types. Generally, obtaining additional text, audio, and tabular data is expensive and time-consuming. There are strong copyright laws protecting text data. Audio files are less common online, and tabular data is primarily from private company databases.

The following section will define the four commonly used data types.

Data input types

The four data input types are self-explanatory, but it is worth clearly defining the data input types and what is out of scope:

- Image definition
- Text definition

- Audio definition

- Tabular data definition

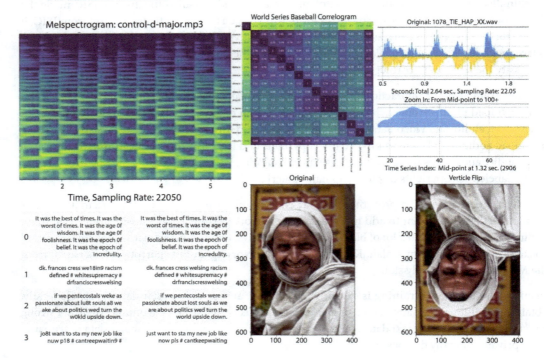

Figure 1.1 – Image, text, tabular, and audio augmentation

Figure 1.1 provides a sneak peek at image, text, tabular and audio augmentation. Later in this book, you will learn how to implement augmentation methods.

Let's get started with images.

Image definition

Image is a large category because you can represent almost anything as an image, such as people, landscapes, animals, plants, and various objects around us. Pictures can also represent action, such as sports, sign language, yoga poses, and many more. One particularly creative use of images is capturing a computer mouse's movement over time to predict whether a user is a computer hacker or not.

The techniques for increasing the number of pictures are horizontal flip, vertical flip, enlarge, zoom in, zoom out, skew, warp, and lighting. Humans are experts at processing images. Thus, if a picture is slightly distorted or darkened, you can still tell that it is the same image. However, this is not the same for a computer. AI represents a color picture as a three-dimensional array of float numbers – the width, height, and RGB as depth. Any image distortion will yield an array with different values.

Graphs, such as time series data charts, and mathematical equation plots, such as 3D topology plots, are outside the scope of image augmentation.

> **Fun fact**
>
> You can eliminate the **overfitting** problem in DL image classification training by creatively using data augmentation methods.

Text augmentation has different concerns than image augmentation. Let's take a look.

Text definition

The primary text input data is in English, but the same techniques for text augmentation can be applied to other West Germanic languages. Python lessons use English as the text input data.

The techniques for supplementing the text input are back translation, easy data augmentation, and albumentation. A few methods might be counterintuitive at first glance, such as deleting or swamping words in a sentence. However, it is an acceptable practice because, in the real world, not everyone writes perfect English.

For example, movie reviewers on the **American Multi-Cinema** (**AMC**) website write incomplete or grammatically incorrect sentences. They omit verbs or use inappropriate words. As a rule of thumb, you should not expect perfect English for text input data in many NLP projects.

If an NLP model is trained in perfect English as text input data, it could cause bias against typical online reviewers. In other words, the NLP model will predict inaccurately when deployed to a real-world audience. For example, in sentiment analysis, the AI system will predict whether a movie review has a positive or negative sentiment. Suppose you trained the system using a perfect English dataset. In that case, the AI system might forecast a **false positive** or **false negative** when people write a short line with misspelled words and grammatical errors.

Language translation, ideograms, and hieroglyphs are outside the scope of this book. Now, let's look at audio augmentation.

Audio definition

Audio input data can be any sound wave recording such as music, speech, and natural sounds. Sound wave attributes such as amplitude and frequency are represented as graphs, which are technically images, but you can't use any image augmentation methods for audio input data.

The techniques for expanding audio input are split into two types: **waveform** and **spectrograph**. For raw audio, the transformation methods range from time-shifting and pitch scaling to random gain, while for spectrographs, the functions are time masking, time stretching, pitch scaling, and many others.

Speech in a language other than English is outside the scope of this book. This is not due to technical difficulties but rather because this book is written in English. Writing about the aftermath effects of switching to a different language would be problematic. Audio augmentation is demanding, but tabular data is even more challenging to expand.

Tabular data definition

Tabular data is information in a relational database, spreadsheet, or text file in **comma-separated values** (**CSV**) format. Tabular data augmentation is a fast-growing field in ML and DL. The tabular data augmentation techniques are transforming, interacting, mapping, and extraction.

Fun challenge

Here is a thought experiment. Can you think of data types other than image, text, audio, and tabular? A hint is *Casablanca* and *Blade Runner*.

There are two parts to this chapter. The first half discussed the various concepts and techniques; what follows is hands-on Python coding on a Python Notebook. The book will use this learn-then-code pattern in all the chapters. It is time to get your hands dirty and write Python code.

Python Notebook

Jupyter Notebook is an open source web application that is the de facto choice for AI, ML, and data scientists. Jupyter Notebook supports multiple computer languages, and the most popular is Python.

Throughout this book, the term **Python Notebook** will be used synonymously for **Jupyter Notebook**, **JupyterLab**, and **Google Colab Jupyter Notebook**.

For Python developers, there are many choices of **integrated development environment** (**IDE**) platforms, such as **Integrated Development and Learning Environment** (**IDLE**), PyCharm, Microsoft Visual Studio, Atom, Sublime, and many more. Still, a Python Notebook is the preferred choice for AI, ML, and data scientists. It is an interactive IDE fit for exploring, coding, and deploying AI projects.

Fun fact

The easiest learning method is reading this book, running the code, and hacking it. This book cannot cover all scenarios; therefore, you must be comfortable with hacking the code so that it matches your real-world dataset. The Python Notebook is designed for interactivity. It gives us the freedom to play, explore, and make mistakes.

Python Notebook is the development tool of choice, and in particular, we will review the following:

- Google Colab
- Python Notebook options
- Installing Python Notebook

Let's begin with Google Colab.

Google Colab

Google Colab Jupyter Notebook with Python is one of the popular options for developing AI and ML projects. All you need is a Gmail account.

Colab can be found at `https://colab.research.google.com/`. The free Colab version is sufficient for the code in this book; the Pro+ version enables more CPU and GPU RAM.

After logging in to Colab, you can retrieve this book's Python Notebooks from the following GitHub URL: `https://github.com/PacktPublishing/data-augmentation-with-python`.

You can start using Colab by using one of the following options:

- The *first method* of opening a Python Notebook is copying it from GitHub. From Colab, go to the **File** menu, choose **Open Notebook**, and then click on the **GitHub** tab. In the **Repository** field, enter the GitHub URL specified previously; refer to *Figure 1.2*. Lastly, select the chapter and Python Notebook (`.ipynb`) file:

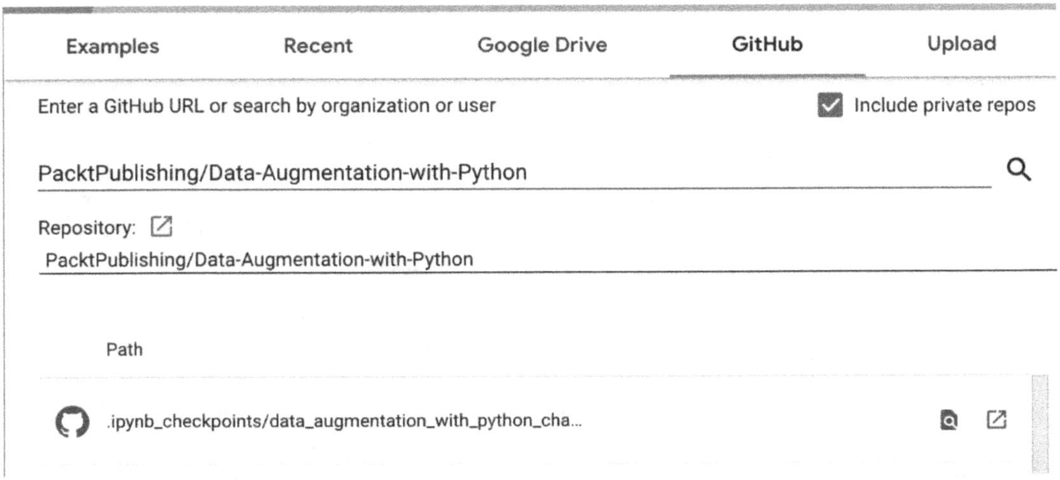

Examples	Recent	Google Drive	GitHub	Upload

Enter a GitHub URL or search by organization or user　☑ Include private repos

PacktPublishing/Data-Augmentation-with-Python　🔍

Repository: ↗
PacktPublishing/Data-Augmentation-with-Python

　　Path

○　.ipynb_checkpoints/data_augmentation_with_python_cha...　🔍 ↗

Figure 1.2 – Loading a Python Notebook from GitHub

- The *second method* of opening a Python Notebook is auto-loading it from GitHub. Go to the GitHub link mentioned previously and click on the Python Notebook (`ipynb`) file. Click the blue-colored **Open in Colab** button, as shown in *Figure 1.3*; it should be on the first line of the Python Notebook. It will launch Colab and load in the Python Notebook automatically:

Figure 1.3 – Loading a Python Notebook from Colab

- Ensure you save a copy of the Python Notebook to your local Google Drive by clicking on the **File** menu and selecting the **Save a copy in Drive** option. Afterward, close the original and use the copy version.

- The *third method* of opening a Python Notebook is by downloading a copy from GitHub. Upload the Python Notebook to Colab by clicking on the **File** menu, choosing **Open Notebook**, then clicking on the **Upload** tab, as shown in *Figure 1.4*:

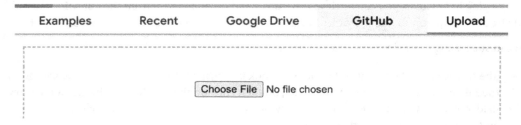

Figure 1.4 – Loading a Python Notebook by uploading it to Colab

> **Fun fact**
>
> For a quick overview of Colab's features, go to `https://colab.research.google.com/notebooks/basic_features_overview.ipynb`. For a tutorial on how to use a Python Notebook, go to `https://colab.research.google.com/github/cs231n/cs231n.github.io/blob/master/jupyter-notebook-tutorial.ipynb`.

Choosing **Colab** follows the same rationale as selecting an IDE: it is based mainly on your preferences. The following section describes additional Python Notebook options.

Additional Python Notebook options

Python notebooks are available in free and paid versions from many online companies, such as Microsoft, Amazon, Kaggle, Paperspace, and others. Using more than one vendor is typical because a Python Notebook behaves the same way across multiple vendors. However, it is similar to choosing an IDE – once selected, we tend to stay in the same environment.

You can use the following feature criteria to select a Python Notebook:

- Easy to set up. Can you load and run a Python Notebook in 15 minutes?
- A free version where you can run the Python Notebooks in this book.
- Free CPU and GPU.
- Free permanent storage for the Python Notebooks and versioning.
- Easy access to GitHub.
- Easy to upload and download the Python Notebooks to and from the local disk drive.
- Option to upgrade to a paid version for faster and additional RAM in terms of CPU and GPU.

The choice of Python Notebook is based on your needs, preferences, or familiarity. You don't have to use Google Colab for the lessons in this book. This book's Python Notebooks will run on, but are not limited to, the following vendors:

- Google Colab
- Kaggle Notebooks
- Deepnote
- Amazon SageMaker Studio Lab
- Paperspace Gradient
- DataCrunch
- Microsoft Notebooks in Visual Studio Code

The cloud-based options depend on having fast internet access at all times, so if internet access is a problem, you might want to install the Python Notebook locally on your laptop/computer. The installation process is straightforward.

Installing Python Notebook

Python Notebook can be installed on a local desktop or laptop for Windows, Mac, and Linux. The advantages of the online version are as follows:

- Fully customizable
- No limit on runtime – that is, no timeout on the Python Notebook during long training sessions
- No rules or arbitrary limitations

The disadvantage is that you have to set up and maintain the environment. For example, you must do the following:

- Install Python and Jupyter Notebook
- Install and configure the NVIDIA graphic card (optional for data augmentation)
- Maintain and update dozens of dependency Python libraries
- Upgrade the disk drive, CPU, and GPU RAM

Installing Python Notebook is easy, requiring just one console or terminal command, but first, check the Python version. Type the following command in the terminal or console application:

```
>python3 --version
```

You should have version 3.7.0 or later. If you don't have Python 3 or have an older version, install Python from https://www.python.org/downloads/.

Install JupyterLab using pip, which contains Python Notebook. On a Windows, Mac, or Linux laptop, use the following command for all three OSs:

```
>pip install jupyterlab
```

If you don't like pip, use conda:

```
>conda install -c conda-forge jupyterlab
```

Other than pip and conda, you can use mamba:

```
>mamba install -c conda-forge jupyterlab
```

Start JupyterLab or Python Notebook with the following command:

```
>jupyter lab
```

The result of installing Python Notebook on a Mac is as follows:

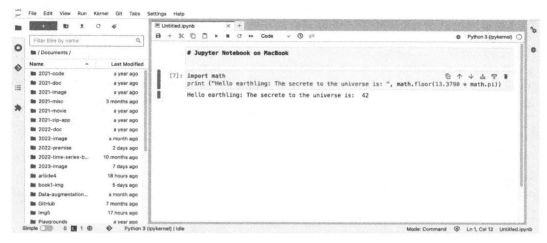

Figure 1.5 – Jupyter Notebook on a local MacBook

The next step is cloning this book's Python Notebook from the respective GitHub link. You can use the GitHub desktop app, the GitHub command on the terminal command line, or the Python Notebook using the magic character exclamation point (!) and standard GitHub command, as follows:

```
url = 'https://github.com/PacktPublishing/Data-Augmentation-with-
Python'
!git clone {url}
```

Regardless of whether you choose the cloud-based options, such as Google Colab or Kaggle, or work offline, the Python Notebook code will work the same. The following section will dive into the Python Notebook programming style and introduce you to Pluto.

Programming styles

The coding style is the standard, tried-and-true method of object-oriented programing and is the variable naming convention for functions and variables.

> **Fun fact**
>
> The majority of Python code you find on blogs and websites is snippets. Therefore, they are not very helpful in studying fundamental topics such as data augmentation. In addition, Python on a Notebook induces lazy practices because programmers think each Notebook's code cell is a separate snippet from the whole. In reality, the entire Python Notebook is one program. Chief among the benefits of using best programming practices is that it's easier to learn and retain knowledge. A programming style may include many standard best practices, but it is also unique to your programming style. Use it to your advantage by learning new concepts and techniques faster, such as how to write data augmentation code.

There are quite a few topics in this section. In particular, we will cover the following concepts:

- Source control
- The `PackTDataAug` class
- Naming convention
- Extend base class
- Referencing library
- Exporting Python code
- Pluto

Let's begin with source control.

Source control

The first rule of programming is to manage the source code version. It will help you answer questions such as, *What did you code last week?*, *What was fixed yesterday?*, *What new feature was added today?*, and *How do I share my code with my team?*

The Git process manages the source code for one person or a team. Among many of Git's virtues is the freedom to make mistakes. In other words, Git allows you to try something new or break the code because you can always roll back to a previous version.

For source control, **GitHub** is a popular website, and **Bitbucket** comes in second place. You can use the Git process from a command-line terminal or Git applications, such as GitHub Desktop.

Google Colab has a built-in Git feature. You have seen how easy it is to load a Python Notebook on Google Colab, and saving it is just as easy. In Git, you must commit and push. The steps are as follows:

1. From the **Colab** menu, click on **File**.
2. Select **Save a copy in GitHub**.

3. Enter your GitHub URL in the **Repository** field and select the code branch.

4. Enter the commit message.

5. Click **OK**:

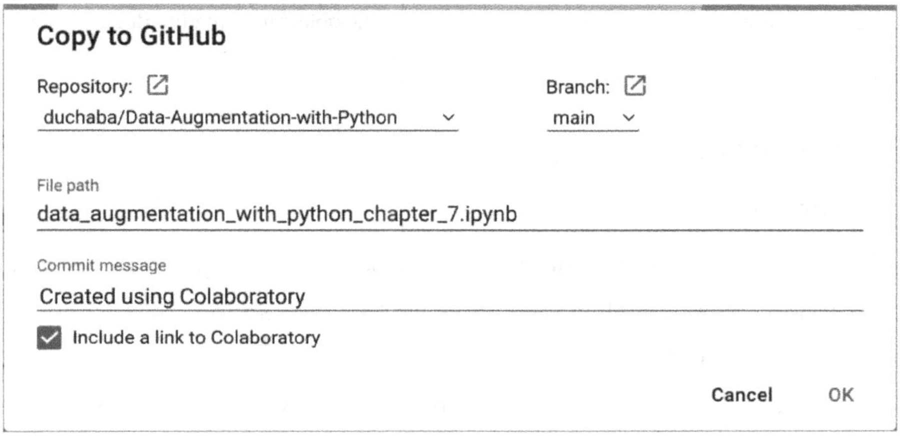

Copy to GitHub

Repository: ↗   Branch: ↗
duchaba/Data-Augmentation-with-Python   ∨   main ∨

File path
data_augmentation_with_python_chapter_7.ipynb

Commit message
Created using Colaboratory

✓ Include a link to Colaboratory

                                              Cancel   OK

Figure 1.6 – Google Colab – saving to GitHub

Figure 1.6 shows the interface between Google Colab Python Notebook and GitHub. Next, we'll look at the base class, PacktDataAug.

The PacktDataAug class

The code for the base class is neither original nor unique to this book. It is standard Python code for constructing an object-oriented class. The name of the object is different for every project. For this book, the name of the class is PacktDataAug.

Every chapter begins with this base class, and we will add new methods to the object using a Python decorator as we learn new concepts and techniques for augmenting data.

This exercise's Python code is in the Python Notebooks and on this book's GitHub repository. Thus, I will not copy or display the complete code in this book. I will show relevant code lines, explain their significance, and rely on you to study the entire code in the Python Notebooks.

The definition of the base class is as follows:

```
# class definition
class PacktDataAug(object):
  def __init__(self,
    name="Pluto",
    is_verbose=True,
    args, **kwargs):
```

PacktDataAug is inherent from the based Object class, and the definition has two optional parameters:

- The name parameter is a string, and it is the name of your object. It has no essential function other than labeling your object.

- is_verbose is a Boolean that tells the object to print the object information during instantiation.

The next topic we will cover is the code naming convention.

Naming convention

The code naming convention is as follows:

- The function's name will begin with an action verb, such as print_, fetch_, or say_.

- A function that returns a Boolean value begins with is_ or has_.

- Variable names begin with a noun, not an action verb.

- There is a heated discussion in the Python community on whether to use camelCase – for example, fetchKaggleData() – or use lowercase with underscores – for example, fetch_kaggle_data(). This book uses lowercase with underscores.

- Functions or variables that begin with underscores are temporary variables or helper functions – for example, _image_auto_id, _drop_images(), and _append_full_path().

- Variable or function abbreviations are sparingly used because the descriptive name is easier to understand. In addition, Colab has auto-complete functionality. Thus, it makes using long, descriptive names easier to type with fewer typos.

The code for instantiating a base class is standard Python code. I used pluto as the object name, but you can choose any name:

```
# Instantiate Pluto
pluto = PackTDataAug("Pluto")
```

The output is as follows:

```
--------------------------------- : ----------------------------
            Hello from class : <class '__main__.PacktDataAug '> Class:
PacktDataAug
                    Code name : Pluto
                    Author is : Duc Haba
--------------------------------- : ----------------------------
```

The base class comes with two simple helper methods. They are both for printing pretty – that is, making the printing of status or output messages neatly centered.

The self._ph() method prints the header line with an equal number of dashes on both sides of the colon character, while the self._pp() function takes two parameters, one for the left-hand side and the other for the right-hand side.

You have already seen the result of instantiating pluto with the default parameter of is_verbose=True. As standard practice, I will not print the complete code in this book. I am relying on you to view and run the code in the Python Notebook, but I will make an exception for this chapter and show you the snippet of code for the is_verbose option. This demonstrates how easy it is to read Python code in the Python Notebook. The snippet is as follows:

```
# code snippet for verbose option
if (is_verbose):
  self._ph()
  self._pp(f"Hello from class {self.__class__} Class: {self.__
class__.__name__}")
  self._pp("Code name", self.name)
  self._pp("Author is", self.author)
  self._ph()
```

> **Fun fact**
> This book's primary goal is to help you write clean and easy-to-understand code and not write compact code that may lead to obfuscation.

Another powerful programming technique is using a Python decorator to extend the base class.

Extend base class

This book has been designed as an interactive journey where you learn and discover new data augmentation concepts and techniques sequentially, from image, text, and audio data to tabular data. The object, pluto, will acquire new methods as the journey progresses. Thus, having a technique to extend the class with new functions is essential. In contrast, providing the fully built class at the beginning of this book would not allow you to embark on the learning journey. Learning by exploration helps you retain knowledge longer compared to learning by memorization.

The @add_method() decorator function extends any class with a new function.

Here is an excellent example of extending the base class. The root cause of Python's most common and frustrating error is having a different library version from the class homework or code snippet copy from the Python community. Python data scientists seldom write code from scratch and rely heavily on existing libraries. Thus, printing the Python library versions on a local or cloud-based server would save hours of aggravating debugging sessions.

To resolve this issue, we can extend the `PacktDataAug` class or use the journey metaphor of teaching Pluto a new trick. The new method, `say_sys_info()`, prints this book's expected system library version on the left-hand side and the actual library version on your local or remote servers on the right-hand side. The decorator's definition for extending the Pluto class is as follows:

```
# using decorator to add new method
@add_method(PackTDataAug)
def say_sys_info(self):
```

After running the aforementioned code cell, you can ask Pluto to print the library version using the following command:

```
# check Python and libraries version
pluto.say_sys_info()
```

The results are as follows:

```
-------------------------------- : ----------------------------
             System time : 2022/07/23 06:36
                Platform : linux
   Pluto Version (Chapter) : 1.0
           Python (3.7.10) : actual: 3.7.12 (default, Apr 24 2022,
17:11:25) [GCC 7.5.0]
          PyTorch (1.11.0) : actual: 1.12.1+cu113
           Pandas (1.3.5) : actual: 1.3.5
              PIL (9.0.0) : actual: 7.1.2
        Matplotlib (3.2.2) : actual: 3.2.2
               CPU count : 2
               CPU speed : NOT available
-------------------------------- : ----------------------------
```

If your result contains libraries that are older versions than this book's expected value, you might run into bugs while working through the lessons. For example, the **Pillow (PIL)** library version is 7.1.2, which is lower than the book's expected version of 9.0.0.

To correct this issue, run the following code line in the Notebook to install the 9.0.0 version:

```
# upgrade to Pillow library version 9.0.0
!pip install Pillow==9.0.0
```

Rerunning `pluto.say_sys_info()` should now show the **PIL** version as 9.0.0.

> **Fun challenge**
>
> Extend Pluto with a new function to display the system's GPU total RAM and available free RAM. The function name can be `fetch_system_gpu_ram()`. A hint is to use the `torch` library and the `torch.to cuda.memory_allocated()` and `torch.cuda.memory_reserved()` functions. You can use this technique to extend any Python library class. For example, to add a new function to the `numpy` library, you can use the `@add_method(numpy)` decorator.

There are a few more programming-style topics. Next, you'll discover how best to reference a library.

Referencing a library

Python is a flexible language when it comes to importing libraries. There are aliases and direct imports. Here are a few examples of importing the same function – that is, `plot()`:

```
# display many options to import a function
from matplotlib.pyplot import plot
import matplotlib.pyplot
import matplotlib.pyplot as plt # most popular
# more exotics importing examples
from matplotlib.pyplot import plot as paint
import matplotlib.pyplot as canvas
from matplotlib import pyplot as plotter
```

The salient point is that all these examples are valid, and that is both good and bad. It enables flexibility, but at the same time, sharing code snippets online or maintaining code can lead to frustration when they break. Python often gives an unintelligible error message when the system cannot locate the function. To fix this bug, you need to know which library to upgrade. The problem is compounded when many libraries use the same function name, such as the `imread()` method, which appears in at least four libraries.

By adhering to this book's programming style, when the `imread()` method fails, you know which library needs to be upgraded or, in rare conditions, downgraded. The code is as follows:

```
# example of use full method name
import matplotlib
matplotlib.pyplot.imread()
```

`matplotlib` might need to be upgraded, or equally, you might be using the wrong `imread()` method. It could be from **OpenCV** version 4.7.0.72. Thus, the call should be `cv2.imread()`.

The next concept is exporting. It may not strictly belong to the programming style, but it is necessary if you wish to reuse and add extra functions to this chapter's code.

Exporting Python code

This book ensures that every chapter has its own Python Notebook. The advanced image, text, and audio chapters need the previous chapter's code. Thus, it is necessary to export the selected Python code cells from the Python Notebook.

The Python Notebook has both markup and code cells, and not all code cells must be exported. You only need to export code cells that define new functions. For the code cells that you want to export to a Python file, use the Python Notebook `%%writefile file_name.py` magic command at the beginning of the code cells and `%%writefile -a file_name.py` to append additional code to the file. `file_name` is the name of the Python file – for example, `pluto_chapter_1.py`.

The last and best part of the programming style is introducing **Pluto** as your coding companion.

Pluto

Pluto uses a whimsical idea of teaching by including dialogs with an imaginary digital character. We can give Pluto tasks to complete. It has a friendly tone, and sometimes the author addresses you directly. It moves away from the direct lecturing format. There are scholarly papers that explain how lecturing in monologue is not the optimal method for learning new concepts, such as the article *Why Students Learn More From Dialogue- Than Monologue-Videos: Analyses of Peer Interactions* by Michelene T. H. Chi, Seokmin Kang, and David L. Yaghmourian that was published by the *Journal of the Learning Sciences* in 2016.

You are most likely reading this book alone rather than engaging in a group, learning how to write augmentation code together. Thus, creating an imaginary companion as the instantiated object might infuse imagination. It makes writing and reading more accessible – for example, the `pluto.fetch_kaggle_data()` function is self-explanatory, and little additional documentation is needed. It simplifies Python code to a common subject and action-verb-sentence format.

Fun challenge

Change the object name from **Pluto** to your favorite canine name, such as **Biggy**, **Sunny**, or **Hanna**. It will make the learning process more personal. For example, change `pluto = PackTDataAug("Pluto")` to `hanna = PackTDataAug("Hanna")`.

Fair warning: Do not choose your beloved cat as the object's name because felines will not listen to any commands. Imagine asking your cat to play fetch.

Summary

In this chapter, you learned that data augmentation is essential for achieving higher accuracy prediction in DL and generative AI. Data augmentation is an economical option for extending a dataset without the difficulty of purchasing and labeling new data.

The four input data types are image, text, audio, and tabular. Each data type faces different challenges, techniques, and limitations. Furthermore, the dataset dictates which functions and parameters are suitable. For example, people's faces and aerial photographs are image datasets, but you can't expand the data by vertically flipping people's images; however, you can vertically flip aerial photos.

In the second part of this chapter, you used **Python notebooks** to reinforce your learning of these augmentation concepts. This involved selecting a Python Notebook as the default IDE to access a cloud-based platform, such as **Google Colab** or **Kaggle**, or installing the Python Notebook locally on your laptop.

The *Programming styles* section lay the foundation for the Python Notebook's structure. It touched on **GitHub** as a form of source control, using base classes, extending base classes, long library function names, exporting to Python, and introducing Pluto.

This chapter laid the foundation with Pluto as the main object. Pluto does not start with complete data augmentation functions – he begins with a minimum structure, and as he learns new data augmentation concepts and techniques from chapter to chapter, he will add new methods to his arsenal.

By the end of this book, Pluto and you will learn techniques regarding how to augment image, text, audio, and tabular data. In other words, you will learn how to write a powerful image, text, audio, and tabular augmentation class from scratch using real-world data, which you can reuse in future data augmentation projects.

Throughout this chapter, there were *fun facts* and *fun challenges*. Pluto hopes you will take advantage of what's been provided and expand your experience beyond the scope of this chapter.

In *Chapter 2, Biases in Data Augmentation*, Pluto and you will explore how data augmentation can increase biases. Using data biases as a guiding principle to data augmentation is an often-overlooked technique.

2
Biases in Data Augmentation

As **artificial intelligence (AI)** becomes embedded in our society, biases in AI systems will adversely affect your quality of life. These AI systems, particularly in **deep learning (DL)** and generative AI, depend on the input data you are using to extend data augmentation.

AI systems rely heavily on data to make decisions, and if the data used to train the system is biased, then the AI system will make unfair decisions. It will lead to the unjust treatment of individuals or groups and perpetuate systemic inequalities. AI plays a decisive role in life-changing decisions, such as how much your monthly mortgage insurance rate is, whether you can be approved for a car loan, your application qualification for a job, who will receive government assistance, how much you pay for milk, what you read on social media newsfeeds, and how much oil your country will import or export, to name a few.

By learning data biases before diving deep into learning data augmentation, you will be able to help develop ethical and fair AI systems that benefit society. It will help you make informed decisions about the data they use and prevent the perpetuation of existing biases and inequalities. Additionally, understanding data bias will help you make informed decisions about the data collection process and ensure it's representative and unbiased.

Data biases may be problematic for data scientists and college students because they are seldom discussed or unavailable in college courses. There is no ready-made fairness matrix to follow programmatically for data augmentation. Maybe by using the latest generative AI, the biases may even originate from computer systems and not be so heavily due to humans.

There are many strategies to provide protected and safe software products and services, but AI systems require new processes and perspectives. Trustworthy and responsible AI is about fairness, ethical design, and minimizing biases. Achieving trustworthy AI starts with transparency, datasets, **test, evaluation, validation, and verification (TEVV)**, as defined by the *Standard for Identifying and Managing Bias in Artificial Intelligence, National Institute of Standards and Technology (NIST)* special publication 1270.

> **Fun fact**
>
> In 2016, Twitter corrupted the Microsoft AI chatbot **Tay** in 1 day. Microsoft created Tay for online casual and playful conversation. Tay was designed to learn and take input from raw, uncurated data and comments from the web. The Twitter community thought it would be fun to teach Tay with misogynistic, racist, and violent tweets. To this day, Tay is a poster child for lessons learned in data bias input for AI. As one blogger put it, *"Flaming garbage pile in, flaming garbage pile out."*

This chapter will provide a crash course on recognizing the differences in **computation**, **human**, and **systemic** biases. We will learn about bias but not practice how to compute bias programmatically. The fairness and confusion matrixes are used to gauge AI's prediction in terms of **true-positive**, **false-positive**, **true-negative**, and **false-negative**. However, the fairness and confusion matrixes are used for building AI systems, not data augmentation. While looking at real-world text datasets, we will attempt to write Python code for a fairness matrix with word counts and misspelled words, but for the most part, we will rely on Pluto and your observations to name the biases in image and text data.

The Python code in this chapter will focus on helping you learn how to download real-world datasets from the *Kaggle* website. The later chapters will reuse the helper and wrapper functions shown in this chapter.

By the end of this chapter, you will have a deeper appreciation for a balanced dataset. In particular, we will cover the following topics:

- Computational biases
- Human biases
- Systemic biases
- Python Notebook
- Image biases
- Text biases

Pluto will begin with the easier of the three biases – computational biases.

Computational biases

Before we start, a fair warning is that you will not be learning how to write Python code to calculate a numeric score for computational bias in datasets. The primary focus of this chapter is to help you learn how to fetch real-world datasets from the Kaggle website and use observation to spot biases in data. There will be some coding to calculate the fairness or balance in the datasets.

For example, we will compute the word counts per record and the misspelled words in the text datasets.

You may think all biases are the same, but it helps to break them into three distinct categories. The bias categories' differences can be subtle when first reading about data biases. One method to help distinguish the differences is to think about how you could remove or reduce the error in AI forecasting. For example, computational biases can be resolved by changing the datasets, while systemic biases can be fixed by changing the deployment and access strategy of the AI system.

Computational biases originate from the unevenness in the dataset for the general population. In other words, it favors or underrepresents one group or data category. The prejudices could be unintentional or deep-seated. The data is skewed higher than the usual randomness. As a result, the algorithm will be plagued with higher false-positive and false-negative predictions.

Dataset representation (DR) and **machine learning algorithms (MLAs)** are two types of computation biases. DR is easier to understand and more closely related to augmenting data. Many of the examples in this section are from DR biases. MLA is specific to a project and can't be generalized.

Here are a few examples of computational biases:

- *Kodak's Shirley Cards Set Photography's Skin-Tone Standard* from the mid-1970s is one of the more famous examples of technology biases. The **Shirley card** from Kodak is used to calibrate the image, such as skin tone and shadow, before printing people's pictures. It is a part of the setting up process and is frequently used at the printing facility. Shirley is the name of an employee at Kodak. Because of this innocent and unintentional discrimination, for three decades, photos printed in the USA did not show the true skin tone of anyone who did not have a white skin tone.

- *Google Open AI DALL-E 2*, from 2022, is an AI model that generates pictures from texts. For example, you can type the input as *a hippo eating broccoli wearing a pink polka dot swimsuit*, and DALL-E 2 will generate the picture. Even with this highly touted technology breakthrough, there are prejudices, as reported by the *NBC Tech* news written by Jake Traylor in the article *No quick fix: How OpenAI's DALL-E 2 illustrated the challenges of bias in AI*. For example, in DALL-E, a builder produced images featuring only men, while the caption of a flight attendant generated only images of women. DALL-E 2 additionally inherits various biases from its training data, and its outputs sometimes reinforce societal stereotypes.

- The United Kingdom's **Information Commissioner's Office (ICO)** disclosed on July 2022 that AI automation's potential discrimination could have grave consequences for society. For example, it could result in unfair or biased job rejection, bank loans, or university acceptance. In addition, coinciding with the ICO is the *Guardian newspaper* article *UK data watchdog investigates whether AI systems show racial bias*, by Dan Milmo. The ICO's goal is to create a fair and ethical AI system guideline.

Figure 2.1 – Generative AI, Stable Diffusion forked

Figure 2.1 displays a hippo eating broccoli. On that fun note, we have concluded this section on computational biases. Pluto is a digital dog but can speak about human biases, which he'll do in the next section.

Human biases

Human biases are even harder to calculate using Python code. There is no Python or other language library for computing a numeric score for human bias in a dataset. We rely on observation to spot such human biases. It is time-consuming to manually study a particular dataset before deriving possible human biases. We could argue that it is not a programmer's or data scientist's job because there is no programable method to follow.

Human biases reflect systematic errors in human thought. In other words, when you develop an AI system, you are limited by the algorithm and data chosen by you. Thus, the prediction of a limited outcome could be biased by your selections. These prejudices are implicit in individuals, groups, institutions, businesses, education, and government.

There is a wide variety of human biases. Cognitive and perceptual biases show themselves in all domains and are not unique to human interactions with AI. There is an entire field of study centered around biases and heuristics in thinking, decision-making, and behavioral economics, such as anchoring bias and confirmation bias.

As data scientists that are augmenting data, by simply being aware of the inherent human prejudices, we can call out the flaws in the data before developing and training the model.

Here are a few examples of human biases in real-world AI systems:

- The **People's Republic of China (PRC)** implemented facial recognition AI in the province of Xinjiang to monitor ethnic minorities such as the Uyghurs. It is the first known example of a government using AI specifically for racial profiling. The system is flawed with discrimination as it identifies the poor and the old as ethnic minorities in false-positive predictions. Compounding the problem is when Myanmar bought the PRC system to crack down on political dissidents, as the **Council on Foreign Relations** reported in their *The Importance of International Norms in Artificial Intelligence Ethics* article in 2022.

- To stop advertisers from abusing the AI Facebook newsfeed, Meta limited the target algorithm from using health, race, ethnicity, political affiliation, religion, and sexual orientation. **NPR** reported in the article that Facebook had scrapped advertised targeting based on politics, race, and other "*sensitive*" topics. The changes took effect on January 10 across Meta's apps, including Facebook, Instagram, Messenger, and the Audience Network. It is reported that advertisers microtargeted people with tailored messages. In other words, the advertisers excluded people based on protected characteristics and targeted advertisements using anti-Semitic phrases.

- The article *Racial Bias Found in a Major Health Care Risk Algorithm*, published by **Scientific American** on October 4, 2019, found many biases in the healthcare system. Black patients would pay more for interventions and emergency visits. In addition to incurring higher costs, black patients would receive lesser-quality care. AI scientists used race and wealth in historical data to train the healthcare system. Thus, the system displayed prejudice toward minority groups and affected 200 million Americans.

Fun challenge

This challenge is a thought experiment. How could you build an AI without biases, given a substantial budget and ample time? Hint: think about when we had world peace or no crime in our city. It can't be an absolute answer. It has to be a level of acceptance.

It may be challenging to see the differences between human and computational biases. Some biases are not one or the other. In other words, they are not mutually exclusive – you can have both human and computational biases in one AI system.

Human biases are difficult to identify because they shape our perception of the world. However, systemic biases may be easier to address in theory, but may be challenging to put into practice.

Systemic biases

If we cannot conceive a method to calculate computational and human biases, then it is impossible to devise an algorithm to compute systemic biases programmatically. We must rely on human judgment to spot the systemic bias in the dataset. Furthermore, it has to be specific to a particular dataset with a distinct AI prediction goal. There are no generalization rules and no fairness matrix to follow.

Systemic biases in AI are the most notorious of all AI biases. Simply put, systemic discrimination is when a business, institution, or government limits access to AI benefits to a group and excludes other underserved groups. It is insidious because it hides behind society's existing rules and norms. Institutional racism and sexism are the most common examples. Another AI accessibility issue in everyday occurrences is limiting or excluding admission to people with disabilities, such as the sight and hearing impaired.

The poor and the underserved have no representation in the process of developing the AI system, but they are forced to accept the AI's prediction or forecast. These AIs make significant life decisions, including those for the poor and underserved, such as how much to pay for a car or health insurance, options for housing or a business bank loan, or whether they are eligible for medical treatments.

Here are a few real-world examples of AI systemic biases:

- The article *Researchers use AI to predict crime, biased policing in major U.S. cities like L.A.*, published by The Los Angeles Times on July 4, 2022, found AI biases in policing crimes. The University of Chicago's AI crime prediction system does not address law enforcement systemic biases. The forecast for possible crime locations, or hot spots, is based on flawed input and environmental factors associated with poor neighborhoods, rather than the actual locations where crimes are committed. The AI reflects the systemic bias in law enforcement practices and procedures. Thus, it forecasts a higher crime rate in poor neighborhoods because of the police's prior systemic biases.

- The US Department of Justice reviewed the **Prisoner Assessment Tool Targeting Estimated Risk and Needs (PATTERN)** and found systemic bias in who can access PATTERN. This was discussed in *Addressing an Algorithmic PATTERN of Bias*, published by the Regulatory Review on May 10, 2020. The report reinforces the Justice Department's biased view that low-risk criminals should be the only ones eligible for early release. PATTERN classifies inmates as low, medium, or high risk, which forecasts if those individuals would engage in crime after release. Since PATTERN is limited to a particular group, it precludes other inmates from early release.

- The article *The Potential For Bias In Machine Learning And Opportunities For Health Insurers To Address It* reports growing concerns about how ML can reflect and perpetuate past and present systemic inequities and biases. This was published by Health Affairs in February 2022 published the article. Limited access to the likelihood of hospitalization, admission to pharmacies, and missing or incomplete data are a few systemic biases for racism and underrepresented populations. Thus, the AI predictions may reflect those systemic biases, and the policy decisions based on the forecast risk reinforcing and exacerbating existing inequities.

> **Fun challenge**
> This challenge is a thought experiment. Which category of biases is easier to spot? Hint: think about company profit.

Computation, human, and systemic biases have similarities and are not mutually exclusive. There is no algorithm or libraries to guide you in coding. It relies on your observation from studying the datasets. At this point, Pluto is ready to learn about fetching real-world datasets from the *Kaggle* website. Optionally, he will ask you to spot the biases in the datasets.

Python Notebook

This chapter's coding lessons primarily focus on downloading real-world datasets from the *Kaggle* website. The later chapters rely on or reuse these fetching functions.

In the previous chapter, you learned about this book's general rules for development on the Python Notebook. The object-oriented class named **Pluto** contains the methods and attributes, and you add new methods to Pluto as you learn new concepts and techniques. Review *Chapter 1* if you are uncertain about the development philosophy.

In this book, the term **Python Notebook** is used synonymously for **Jupyter Notebook**, **JupyterLab**, and **Google Colab Notebook**.

> **Fun challenge**
> Pluto challenges you to change the object's name from **Pluto** to any other name. If you do change the name, then substitute that name where you see **Pluto** in your text and code. For example, if you change the object name to **Sandy**, then `pluto.draw_batch_image()` becomes `sandy.draw_batch_image()`.

Starting with this chapter, the setup process for using the Python Notebook will be the same for every chapter. The goal of this chapter is to help you gain a deeper understanding of the datasets and not to write Python code for calculating the bias value for each dataset. The setup steps are as follows:

1. Load Python Notebook.
2. Clone GitHub.
3. Instantiate Pluto.
4. Verify Pluto.
5. Create Kaggle ID.
6. Download real-world datasets.

Let's start with loading the Python Notebook.

Python Notebook

The first step is to locate the `data_augmentation_with_python_chapter_2.ipynb` file. It is in this book's GitHub repository at `https://github.com/PacktPublishing/Data-Augmentation-with-Python`. Refer to *Chapter 1* if you forgot how to load the Python Notebook.

The next step is to clone the GitHub repository.

GitHub

The second step is locating the `~/Data-Augmentation-with-Python/pluto/pluto_chapter_1.py` file. It's in the main GitHub repository for this book, under the `pluto` folder.

Pluto is using the Python Notebook on **Google Colab**. It starts with a new session every time – that is, no permanent storage is saved from the previous session. Thus, the faster and easier method to load all the required files is to clone the GitHub repository. It could be this book's GitGub or the GitHub repository that you forked.

From this point onward, all commands, code, and references are from the `data_augmentation_with_python_chapter_2.ipynb` Python Notebook.

Pluto uses the `!git clone {url}` command to clone a GitHub repository, where `{url}` is the link for the GitHub repository. The code snippet is as follows:

```
# load from official GitHub repo.
!git clone https://github.com/PacktPublishing/Data-Augmentation-with-Python
# optional step, sustitute duchaba with your GitHub space
!git clone https://github.com/duchaba/Data-Augmentation-with-Python
```

In the Python Notebook, any code cell that begins with an exclamation point (!) will tell the system to run as a system shell command line. For Google Colab, it is a **bsh** shell. In addition, all code that begins with a percent sign (%) are special commands. They are called **magic keywords** or **magic commands**.

> **Fun fact**
>
> Jupyter Notebook's built-in magic commands provide convenience functions to the underlying **operating system** (**OS**) kernel. The magic commands begin with the percent sign character (%). For example, `%ldir` is for listing the current directory files, `%cp` is for copying files in your local directory, `%debug` is for debugging, and so on. The helpful `%lsmagic` command is for listing all the available magic commands supported by your current Python Notebook environment. The exclamation character (!) is for running the underlying OS command-line function. For example, in a Linux system, `!ls -la` is for listing the files in the current directory, while `!pip` is for installing Python libraries.

Now that you have downloaded the Pluto Python code, the next step is to instantiate Pluto.

Pluto

The Python Notebook's magic command for instantiating Pluto is as follows:

```
# instantiate pluto
pluto_file='Data-Augmentation-with-Python/pluto/pluto_chapter_1.py'
%run {pluto_file}
```

The output is as follows:

```
--------------------------- : ---------------------------
        Hello from class : <class '__main__.PackTDataAug'> Class:
PackTDataAug
            Code name : Pluto
            Author is : Duc Haba
--------------------------- : ---------------------------
```

The next-to-last step in the setup process is to verify that Pluto is running with the correct version.

Verifying Pluto

For double-checking, Pluto runs the following function:

```
# Are you ready to play?
pluto.say_sys_info()
```

The results should be similar to the following output:

```
--------------------------- : ---------------------------
            System time : 2022/08/16 06:26
               Platform : linux
    Python version (3.7+) : 3.7.13 (default, Apr 24 2022, 01:04:09)
[GCC 7.5.0]
    Pluto Version (Chapter) : 1.0
        PyTorch (1.11.0) : actual: 1.12.1+cu113
         Pandas (1.3.5) : actual: 1.3.5
            PIL (9.0.0) : actual: 7.1.2
     Matplotlib (3.2.2) : actual: 3.2.2
              CPU count : 2
              CPU speed : NOT available
--------------------------- : ---------------------------
```

Since this code is from *Chapter 1*, the Pluto version is **1.0**. Before Pluto can download the dataset from the *Kaggle* website, he needs Kaggle's **key** and **access token**.

Kaggle ID

Pluto uses Kaggle datasets because he wants to learn how to retrieve real-world data for learning data augmentation. It is more impactful than using a small set of dummy data. Thus, the first two steps are installing the library to aid in downloading the Kaggle data and signing up with `Kaggle.com`.

The code for installing and importing can be found in the open source **opendatasets** library by **Jovian**. The function code is in the Python Notebook; here is a code snippet from it:

```
# install opendatasets library
!pip install opendatasets --upgrade
import opendatasets
```

After you create an account on `Kaggle.com`, you will have a **Kaggle username** and receive a **Kaggle key**. Next, go to the **Account** page, scroll down to the **API** section, and click on the **Create New API Token** button to generate the **Kaggle key**:

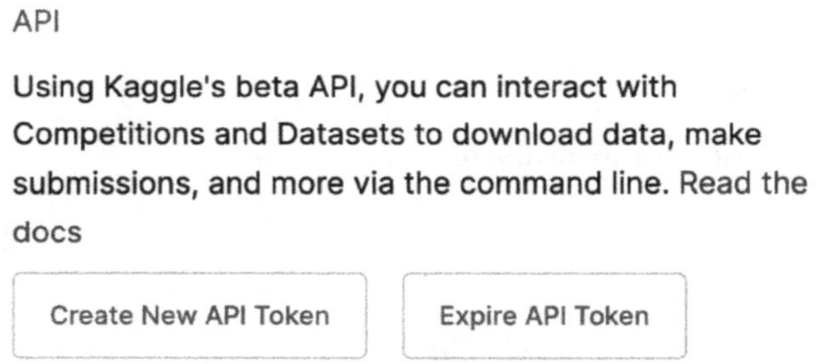

API

Using Kaggle's beta API, you can interact with Competitions and Datasets to download data, make submissions, and more via the command line. Read the docs

| Create New API Token | Expire API Token |

Figure 2.2 – Kaggle Account page – new token

Once you have a **Kaggle username and key**, as shown in *Figure 2.2*, use Pluto's `remember_kaggle_access_key()` wrapper method to store the attributes inside the object. The code uses the Python `self` keyword to store this information – for example, `self.kaggle_username`. The method's definition is as follows:

```
# method definition
def remember_kaggle_access_keys(self,username,key):
```

Other methods will use these attributes automatically. Pluto runs the following method to remember your Kaggle username and key:

```
# save Kaggle username and key
pluto.remember_kaggle_access_keys("your_username_here",
  "your_key_here")
```

The _write_kaggle_credit() method writes your Kaggle username and key in two locations – ~/.kaggle/kaggle.json and ./kaggle.json. It also changes the file attribute to 0o600. This function begins with an underscore; hence, it is a helper function used primarily by other methods.

There are two methods for Pluto to fetch data from Kaggle: fetch_kaggle_comp_data(competition_name), where competition_name is the title of the contest, and fetch_kaggle_dataset(url), where url is the link to the dataset.

In the fetch_kaggle_comp_data() wrapper method, the primary code line that does most of the work is as follows:

```
# code snippet for fetcing competition data
kaggle.api.competition_download_cli(str(path))
```

In the fetch_kaggle_dataset() method, the primary code line that does most of the work is as follows:

```
# fetching real-world dataset for the Kaggle website
opendatasets.download(url,data_dir=dest)
```

> **Fun fact**
>
> As of 2022, there are over 2,500 past and current competitions on the Kaggle website and more than 150,000 datasets. These datasets are diverse, from medical and financial to other industry-specific datasets.

Image biases

Pluto has access to thousands of datasets, and downloading these datasets is as simple as replacing the **URL**. In particular, he will download the following datasets:

- The *State Farm distracted drivers detection (SFDDD)* dataset
- The *Nike shoes* dataset
- The *Grapevine leaves* dataset

Let's start with the SFDDD dataset.

State Farm distracted drivers detection

To start, Pluto will slow down and explain every step in downloading the real-world datasets, even though he will use a wrapper function, which seems deceptively simple. Pluto will not write any Python code for programmatically computing the bias fairness matrix values. He relies on your observation to spot the biases in the dataset.

Give Pluto a command to fetch, and he will download and **unzip** or **untar** the data to your local disk space. For example, in retrieving data from a competition, ask Pluto to fetch it with the following command:

```
# fetch real-world data
pluto.fetch_kaggle_comp_data(
  "state-farm-distracted-driver-detection")
```

Since this data is from a competition, you must join the State Farm competition before downloading the dataset. You should go to the **State Farm Distracted Driver Detection** competition and click the **Join** button. The description for the competition from the *Kaggle* website is as follows:

> *"State Farm hopes to improve these alarming statistics and better insure their customers by testing whether dashboard cameras can automatically detect drivers engaging in distracting behaviors. Given a dataset of 2D dashboard camera images, State Farm is challenging Kagglers to classify each driver's behavior."*

State Farm provided the dataset, announced in 2016. The rules and usage licenses can be found at https://www.kaggle.com/competitions/state-farm-distracted-driver-detection/rules.

Fun fact

You have to join a Kaggle competition to download competition data, but you don't need to enter a competition to download a Kaggle dataset.

Not all the methods are in the Python Notebook's **global space** but in the Pluto object. Hence, you can't access a wrapper function directly. You have to prefix it with `pluto`. For example, you can't do the following:

```
# example of wrong syntax

fetch_kaggle_dataset(url)
```

However, using the `pluto` prefix is correct, as shown here:

```
# example of correct syntax
pluto.fetch_kaggle_dataset(url)
```

Before Pluto displays the image in batches, he must write a few simple code lines to check if the downloads are correct:

```
# read the image file
f = 'state-farm-distracted-driver-detection/imgs/train/c0/img_100026.jpg'
img = PIL.Image.open(f)
# display image using Python Notebook build-in command
display(img)
```

The output is as follows:

Figure 2.3 – State Farm Distracted driver

The SFDDD dataset consists of 22,423 images, and viewing one photo at a time, as shown in *Figure 2.3*, will not help Pluto to see the biases. Pluto loves putting lists and tabular data into the Python pandas library. Luckily, the State Farm competition comes with a **comma-separated values (CSV)** file. It will make writing the fetch_df(self, csv) method easier. The relevant line of code is as follows:

```
# code snippet to import into Pandas
df = pandas.read_csv(csv)
```

Pluto uses the fetch_df(self, csv) wrapper function to download the data, and he uses Pandas to display the last three rows. The code is as follows:

```
# fetch data
pluto.df_sf_data = pluto.fetch_df('state-farm-distracted-driver-
detection/driver_imgs_list.csv')
# display last three records
pluto.df_sf_data.tail(3)
```

The result is as follows:

	subject	classname	img
22421	p081	c9	img_25946.jpg
22422	p081	c9	img_67850.jpg
22423	p081	c9	img_9684.jpg

Figure 2.4 – State Farm data – last three rows

Pluto likes the data in the original CSV file, shown in *Figure 2.4*, but it does not have a column with a full path to an image file. Pandas makes creating a new column containing the full image path super easy. There are no complicated **for loops** or **if else** statements. There are only two lines of code for the wrapper function, `build_sf_fname(self, df)`, where `df` is the original DataFrame. The code snippet is as follows:

```
# code snippet to create full image path
root = 'state-farm-distracted-driver-detection/imgs/train/'
df["fname"] = f'{root}/{df.classname}/{df.img}'
```

The full function code can be found in the Python Notebook. Pluto adds the full path name column and displays the first three rows with the following code:

```
#create new fname column
pluto.build_sf_fname(pluto.df_sf_data)
pluto.df_sf_data.head(3)
```

The result is as follows:

	subject	classname	img	fname
0	p002	c0	img_44733.jpg	state-farm-distracted-driver-detection/imgs/train/c0/img_44733.jpg
1	p002	c0	img_72999.jpg	state-farm-distracted-driver-detection/imgs/train/c0/img_72999.jpg
2	p002	c0	img_25094.jpg	state-farm-distracted-driver-detection/imgs/train/c0/img_25094.jpg

Figure 2.5 – State Farm data – full path image name

For double-checking, Pluto writes a few lines of simple code to display an image from the pandas `fname` column, as shown in *Figure 2.5*, using the **PIL** library. The code is as follows:

```
# display the image
img = PIL.Image.open(pluto.df_sf_data.fname[0])
display(img)
```

The resulting image is as follows:

Figure 2.6 – State Farm data – the fname column

Figure 2.6 shows a driver. Using the fname column, drawing a batch or collection of images is relatively easy. The draw_batch() wrapper function's definition is as follows:

```
# function definition
def draw_batch(self, df_filenames,
    disp_max=10,
    is_shuffle=False,
    figsize=(16,8)):
```

= df_filenames is the list of file =names, and it is in a pandas DataFrame. disp_max defaults to 10, which is an increment of 5, as in five photos per row. is_shuffle defaults to False. If you can set it to True, each batch is randomly selected. Lastly, figsize is the size of the output from the **Matplotlib** library, where the first number is the **width** and the second number is the **height**. The default is (16,8).

Using the draw_batch() wrapper method, Pluto can draw any photo collection. For example, Pluto can draw 10 random images from the SFDDD competition with the following code:

```
# display image batch
x = pluto.draw_batch(pluto.df_sf_data["fname"],
    is_shuffle=True)
```

The result is as follows:

Figure 2.7 – State Farm data – draw_patch()

Pluto runs the code repeatedly to see different images in the dataset, as shown in *Figure 2.7*. For example, he can draw 20 random images at a time using the following code:

```
# display image batch
x = pluto.draw_batch(pluto.df_sf_data["fname"],
  is_shuffle=True,
  disp_max=20,
  figsize=(18,14))
```

The output is as follows:

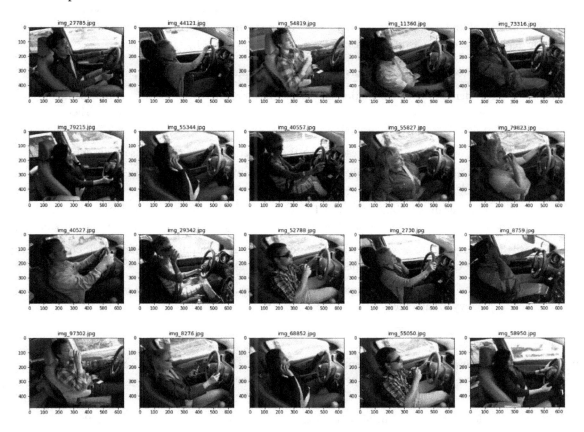

Figure 2.8 – State Farm data – 20 randomly selected images

Figure 2.8 displays 20 photos of drivers. Using the `fetch_kaggle_comp_data()`, `fetch_df()`, and `draw_batch()` wrapper functions, Pluto can retrieve any of the thousand real-world datasets from Kaggle.

Fun challenge

This challenge is a thought experiment. Before reading Pluto's answer, what biases do you see in the images? It is optional, and there is no algorithm or library that you can use to compute the bias fairness value. It relies on your observation.

Pluto read the SFDDD's goal and thought about computational, human, and systemic biases. The following bullet points are not errors to be fixed, but they could be biases. These biases are observations from *Figure 2.7* of underrepresented groups. Pluto assumes the long-term goal of the SFDDD is for it to be deployed across the United States:

- Pluto does not see any older adults as drivers in the dataset.
- The driver demographic distribution is limited. There are about a dozen drivers represented in the dataset, and the long-term goal is to deploy this AI system in the United States. Therefore, the AI system will be trained on a limited number of drivers.
- There are few vehicle types represented in the dataset. They are primary sedans, compacts, or SUVs. A sports car or truck interior is different, which might affect the prediction of false positives or false negatives.
- There are other distracting activities while driving that are not represented, such as eating ice cream, watching an event unfolding outside of the car, head or hair grooming, and so on.
- All drivers in the dataset wear urban-style clothing. More elaborate or ethnic-centric clothing styles might cause the AI to predict false positives or false negatives.
- The goal is to save lives. Thus, a systemic bias could be affordable access to everyone, not just the tech-savvy urban elites.

> **Fun challenge**
> This challenge is a thought experiment. Can you find other biases? There are no absolute right or wrong answers. The biases listed here can't be spotted programmatically.

That was a detailed discussion of the SFDDD dataset. Pluto will fetch another dataset from the *Kaggle* website, the *Nike shoes* dataset.

Nike shoes

The Nike shoes dataset was chosen because it will show different biases. Like the State Farm photos, there is no algorithm or library to compute the fairness matrix. We rely on Pluto and your observations.

The *Nike, Adidas, and Converse Shoes Images* (Nike) dataset contains images in folders; there is no CSV file. The Nike dataset's description on the Kaggle website is as follows:

"This dataset is ideal for performing multiclass classification with deep neural networks such as CNNs or simpler machine learning classification models. You can use TensorFlow, its high-level API Keras, sklearn, PyTorch, or other deep/machine learning libraries."

The author is *Iron486*, and the license is **CC0: Public Domain**: `https://creativecommons.org/publicdomain/zero/1.0/`.

Since there is no CSV file for Pluto to import into pandas, Pluto has written the `build_df_fname(self, start_path)` method, where `start_path` is the directory where the data is stored.

The key code line is the `os.walk()` function:

```
# code snippet for generating meta data
for root, dirs, files in os.walk(start_path, topdown=False):
    for name in files:
```

Pluto will perform the three familiar steps for reviewing the Nike dataset. They are as follows:

```
# 1. fetch data
fname='https://www.kaggle.com/datasets/die9origephit/nike-adidas-and-
converse-imaged'
pluto.fetch_kaggle_dataset(fname)
# 2. import meta data to Pandas
pluto.df_shoe_data = pluto.build_shoe_fname(
   'kaggle/nike-adidas-and-converse-imaged/train')
# 3. display image batch
x = pluto.draw_batch(pluto.df_shoe_data["fname"],
   is_shuffle=True,
   disp_max=20,
   figsize=(18,14))
```

The output is as follows:

Figure 2.9 – Nike data – 20 randomly selected images

The following is Pluto's list of data biases observations from *Figure 2.9*:

- The shoes are too clean. Where are the muddy or dirty shoes?
- The photos are professionally taken. Thus, when the AI-powered app is deployed, people might find their app giving a wrong prediction because their pictures are taken haphazardly.
- There is a lack of shoe images in urban, farming, or hiking settings.

Let's ask Pluto to grab one more image dataset before switching gears and digging into the text dataset.

Grapevine leaves

The Grapevine leaves dataset is the third and last example of a real-world image dataset Pluto will fetch from the Kaggle website. The primary goal is for you to practice downloading datasets and importing the metadata into pandas. Incidentally, Pluto will use the Grapevine leaves dataset to name other types of data biases through observation. He does not rely on defining a fairness matrix through coding because it not yet feasible. Maybe the next level of generative AI will be able to process all the photos in a dataset and deduce the biases.

Here is an excerpt from the Grapevine leaves dataset:

> *"The main product of grapevines is grapes that are consumed fresh or processed. In addition, grapevine leaves are harvested once a year as a by-product. The species of grapevine leaves are important in terms of price and taste."*

The authors are *Koklu M., Unlersen M. F., Ozkan I. A., Aslan M. F., and Sabanci K.*, and the license is **CC0: Public Domain**: https://creativecommons.org/publicdomain/zero/1.0/.

The filenames in the Grapevine dataset contains a space character in the filename, which may confuse many Python libraries. Thus, Pluto runs a few simple Linux scripts to convert the space into an underscore. The code snippet is as follows:

```
# remove white space from file and directory name
f2 = 'kaggle/grapevine-leaves-image-dataset/Grapevine_Leaves_Image_
Dataset'
!find {f2} -name "* *" -type f | rename 's/ /_/g'
```

After cleaning up the filenames, Pluto will perform the three familiar steps for fetching, importing, and displaying the Grapevine dataset. The images are in the same folder structure as the Nike photos. Thus, Pluto reuses the same `pluto.fetch_df()` method:

```
# fetch data
fname=' https://www.kaggle.com/datasets/muratkokludataset/grapevine-
leaves-image-dataset'
pluto.fetch_kaggle_dataset(fname)
# import to Pandas
pluto.df_grapevine_data=pluto.fetch_df("kaggle/grapevine-leaves-image-
dataset/Grapevine_Leaves_Image_Dataset")
# display image batch
x = pluto.draw_batch(pluto.df_grapevine_data["fname"],
   is_shuffle=True,
   disp_max=20,
   figsize=(18,14))
```

The output is as follows:

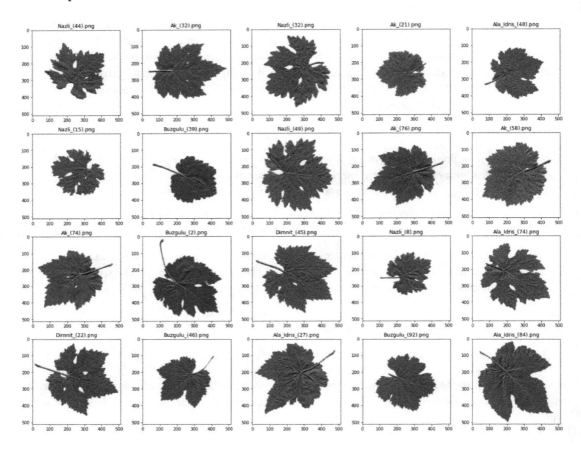

Figure 2.10 – Grapevine data – 20 randomly selected images

The following is Pluto's list of data biases from *Figure 2.10*:

- The photos are too perfect, and undoubtedly, they are uncomplicated to augment and train, but how does the general public use the AI system? If the winemakers access the AI system through an iPhone, the grapevine leaf pictures they take are nothing like the flawless photos in the dataset. The resulting predictions could be false positives.

- Similar to the perfect photo bias, the leaf is flat, and the background is white, which is not common in real-world usage. The training cycle will achieve high accuracy, but it is unsuitable for real-world use.

- If the model is trained as-is and deployed, then the resulting AI will have a systemic bias, only being available for lab technicians and not farmers.

> **Fun challenge**
> There are thousands of image datasets on the *Kaggle* website. Pluto challenges you to select, download, display, and list the biases for three different image datasets.

Other than Distracted Drivers, Nike shoes, and Grapevine Leaves, there are more examples in the Python Notebook. However, next, Pluto will move on from biases to text augmentation.

Text biases

By now, you should recognize the patterns for fetching real-world image datasets and importing metadata into pandas. It is the same pattern for text datasets. Pluto will guide you through two sessions and use his power of observation to name the biases. He could employ the latest in generative AI such as OpenAI GPT3 or GPT4 to list the biases in the text. Maybe he will do that later, but for now, he will use his noggin. Nevertheless, Pluto will attempt to write Python code to gain insight into the texts' structures, such as the word count and misspelled words. It is not the fairness matrix but a step in the right direction.

Pluto searches the Kaggle website for the **Natural Language Processing** (**NLP**) dataset, and the result consists of over 2,000 datasets. He chooses the *Netflix Shows* and the *Amazon Reviews* datasets. Retrieving and viewing the NLP dataset follows the same fetching, importing, and printing steps outlined in the image dataset.

Let's start with the Netflix data.

Netflix

Pluto reuses the wrapper function to download the data. The command is as follows:

```
# fectch real-world dataset
fname='https://www.kaggle.com/datasets/infamouscoder/dataset-Netflix-
shows'
pluto.fetch_kaggle_dataset(fname)
```

The Netflix dataset's description from the Kaggle website is as follows:

> *"The raw data is web scrapped through Selenium. It contains unlabelled text data*
> *of around 9,000 Netflix shows and movies, along with full details such as cast,*
> *release year, rating, description, and so on."*

The author is *InFamousCoder*, and the license is **CC0: Public Domain**: https://creativecommons. org/publicdomain/zero/1.0/.

The second step is to import the data into a pandas DataFrame. The Netflix data comes with a **CSV** file; therefore, Pluto reuses the `fetch_df()` method to import the Netflix reviews into the DataFrame and displays the first three rows, as follows:

```
# import metadata into Pandas
pluto.df_netflix_data = pluto.fetch_df(
    'kaggle/dataset-Netflix-shows/Netflix_titles.csv')
pluto.df_netflix_data[['show_id',
    'type', 'title', 'director', 'cast']].head(3)
```

The result is as follows:

	show_id	type	title	director	cast
0	s1	Movie	Dick Johnson Is Dead	Kirsten Johnson	nan
1	s2	TV Show	Blood & Water	nan	Ama Qamata, Khosi Ngema, Gail Mabalane, Thabang Molaba, Dillon Windvogel, Natasha Thahane, Arno Greeff, Xolile Tshabalala, Getmore Sithole, Cindy Mahlangu, Ryle De Morny, Greteli Fincham, Sello Maake Ka-Ncube, Odwa Gwanya, Mekaila Mathys, Sandi Schultz, Duane Williams, Shamilla Miller, Patrick Mofokeng
2	s3	TV Show	Ganglands	Julien Leclercq	Sami Bouajila, Tracy Gotoas, Samuel Jouy, Nabiha Akkari, Sofia Lesaffre, Salim Kechiouche, Noureddine Farihi, Geert Van Rampelberg, Bakary Diombera

Figure 2.11 – Netflix data, left columns

Figure 2.11 displays the Netflix metadata. The first two steps do not require Pluto to write new code, but Pluto has to write code for the third step, which is to display the movie's title and description. The goal is for Pluto to find any biases in the movie description.

Pandas made writing the `display_batch_text()` wrapper method effortless. The method has no **loops**, **index counter**, **shuffle algorithm**, **or if-else** statements. There are just three lines of code, so Pluto displays the code in its entirety here:

```
# define wrapper function
def print_batch_text(self, df_orig,
    disp_max=10,
```

```
cols= ["title", "description"]):
df = df_orig[cols]
with pandas.option_context("display.max_colwidth", None):
  display(df.sample(disp_max))
return
```

Pluto displays the Netflix movies' titles and descriptions in batch using the following code:

```
# print text batch
pluto.print_batch_text(pluto.df_netflix_data)
```

The result is as follows:

	title	description
5342	Alan SaldaÃ±a: Mi vida de pobre	Mexican comic Alan SaldaÃ±a has fun with everything from the pressure of sitting in an exit row to maxing out his credit card in this stand-up special.
6991	Honey: Rise Up and Dance	Despite discouragement from her loved ones, a talented street dancer tries out for an exclusive dance crew in a bid to win a college scholarship.
5023	Derren Brown: The Push	Mentalist Derren Brown engineers an audacious social experiment demonstrating how manipulation can lead an ordinary person to commit an appalling act.
2143	Rob Schneider: Asian Momma, Mexican Kids	Former "Saturday Night Live" star Rob Schneider returns to the stage and shares his take on life, love and dinosaur dreams in this stand-up special.
7182	Khaani	After a rich politician's son kills a young woman's brother, an unlikely romantic connection complicates her pursuit of justice.
280	Bake Squad	Expert bakers elevate desserts with next-level ideas and epic execution. Now the battle is on to win over clients in need of very special sweets.

Figure 2.12 – Netflix movie title and description

Fun fact

Every time Pluto runs the `print_batch_text()` wrapper function, movie titles and descriptions are displayed. It would be best to run the wrapper function repeatedly to gain more insight into the data.

Figure 2.12 displays a text batch. Pluto has read hundreds of movie descriptions and found no apparent bias. It is a job for a linguist. In general, the English language can have the following biases:

- Religious bias
- Gender bias
- Ethnicity bias
- Racial bias
- Age bias
- Mental health bias
- Former felon bias
- Elitism bias
- LGBTQ bias
- Disability bias

Pluto is not a linguist, but there are other data attributes could contribute to language biases, such as word count and misspelled words. In other words, are the Netflix movie descriptions all relatively the same length? And are there many misspelled words?

This is an attempt to code a small fraction of the fairness matrix. When using a pandas DataFrame, the count_words() method has one line of code. It is as follows:

```
# attempt at fairness matrix, count words
def count_word(self, df, col_dest="description"):
  df['wordc'] = df[col_dest].apply(lambda x: len(x.split()))
  return
```

Pluto counted the number of words in the Netflix movie and double-checked the result by using the following code:

```
# count words and dislay result
pluto.count_word(pluto.df_netflix_data)
pluto.print_batch_text(pluto.df_netflix_data,
  cols=['description','wordc'])
```

The result is as follows:

	description	wordc
6954	When best friends break a blood oath, one of them is cast away to hell, and the other two must save him from the misfits of the underworld.	28
4619	In the grim Alaskan winter, a naturalist hunts for wolves blamed for killing a local boy, but he soon finds himself swept into a chilling mystery.	26
4832	Do-gooding True and Bartleby will go anywhere to make wishes come true â€ from the bottom of the Living Sea to the tip of Mount Tippy Tippy Top!	28
3958	Decades ago, a hero from the stars left this world in peace. Now, the son of Ultraman must rise to protect the Earth from a new alien threat.	28
7344	Sparks fly and romance blooms when Shelby, a vegan chef, meets Greg, a developer â€ until their opposing views clash.	20
5058	In this musical comedy, two rebellious teen girls who love electronic music have a life-changing brush with the divine at a camp run by nuns.	25

Figure 2.13 – Movie description word count

Figure 2.13 displays the word count for each record. The next step is to plot the word count using the **BoxPlot** and **Histogram** graphs. When using a pandas DataFrame, drawing graphs is relatively easy. The two key code lines in the draw_word_count() function are as follows:

```
# code snippet for draw word count
df.boxplot(ax=pic[0],
  column=[wc],
  vert=False,
  color="black")
df[wc].hist(ax=pic[1],
  color="cornflowerblue",
  alpha=0.9)
```

The full function code can be found in the Python Notebook. Pluto draws the BoxPlot and Histogram graphs with the following code:

```
# draw word count
pluto.draw_word_count(pluto.df_netflix_data)
```

The result is as follows:

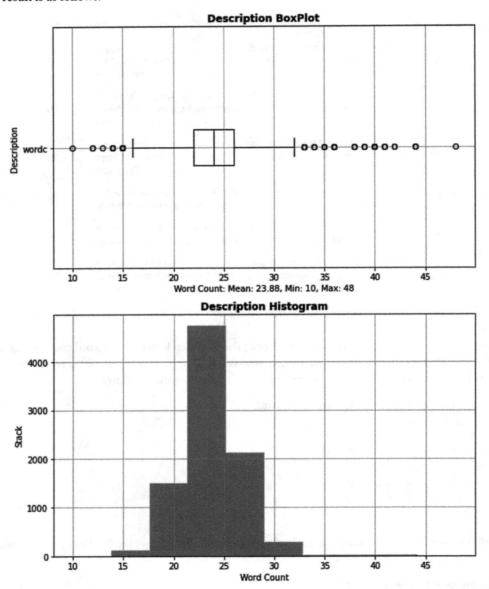

Figure 2.14 – Netflix movie description word count

As shown in *Figure 2.14*, the BoxPlot and Histogram plots show that the distribution is even. There are a few outliers, the mean is 23.88, and the bulk of the Netflix movie descriptions are between 22 and 25 words. Thus, there is no bias here. Pluto investigates the misspelled words next.

Pluto uses the `pip` command to install the **pyspellchecker** library and import the `spellchecker` class. The `check_spelling()` method takes the pandas DataFrame and the designated column as parameters. The function key code lines are as follows:

```
# code snippet for check spelling
df["misspelled"] = df[col_dest].apply(
  lambda x: spell.unknown(self._strip_punc(x).split()))
df["misspelled_count"] = df["misspelled"].apply(
  lambda x: len(x))
```

Pluto checks the Netflix movie descriptions' spelling and uses the `print_batch_text()` function to display the result. The code is as follows:

```
# check spelling
pluto.check_spelling(pluto.df_netflix_data)
# print batch text withh correct spelling
pluto.print_batch_text(pluto.df_netflix_data,
  cols=['description', 'misspelled'])
```

The result is as follows:

	description	misspelled
884	Finding life in all thatâ€ s left behind, a detail-oriented trauma cleaner and his estranged uncle deliver untold stories of the departed to loved ones.	{'thatâs', 'detailoriented'}
8301	Following a friend's suicide after her husband dumps her for a younger model, three women plot payback against their two-timing exes.	{'twotiming'}
3571	Elite street racers from around the world test their limits in supercharged custom cars on the biggest, baddest automotive obstacle course ever built.	set()
8758	From Moscow to Mexico City, three BBC journalists delve into the inner workings of some of the most burgeoning metropolises on Earth.	{'bbc', 'metropolises'}
7143	Juana InÃ©s de la Cruz, a powerful feminist nun involved in a forbidden love affair with a woman, faces oppression in 17th-century Mexico.	{'inãs', '17thcentury', 'juana'}
5534	A famous designer who's always pretended to be gay finds himself in crisis mode when threatened with exposure as the woman-chasing straight guy he is.	{'womanchasing'}

Figure 2.15 – Netflix misspelled words

Figure 2.15 displays the misspelled words. Pluto displays this data in graphs by reusing the same `draw_word_count ()` function, as follows:

```
# draw word count
pluto.draw_word_count(pluto.df_netflix_data,
  wc='misspelled_count')
```

The result is as follows:

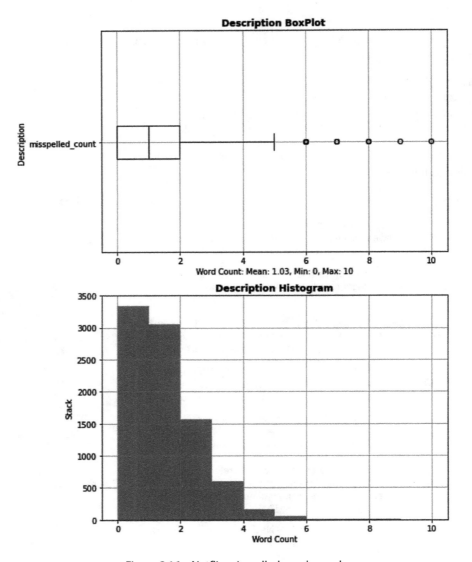

Figure 2.16 – Netflix misspelled words graph

The misspelled words are mostly person or product names, as shown in *Figure 2.16*. The average is 0.92 per Netflix movie description and there are only a handful of outliners. Without a linguist's help, Pluto can't find any biases in the Netflix movie description. Let's move on to the Amazon reviews and see if we can find any biases.

Amazon reviews

The Amazon reviews dataset is the last real-world text dataset to download for this chapter. Pluto follows the same pattern, and you should now be comfortable with the code and ready to download any real-world datasets from the Kaggle website. In addition, as with the Netflix data, Pluto will use his powerful insight, as a digital dog, to find the biases in the text. He will use the same techniques and library to programmatically find the word count and misspelled words.

Pluto will not explain how the code is written for the Amazon reviews because he re-used the same functions in the Netflix data. The complete code can be found in the Python Notebook. The bare code snippet is as follows:

```
# fetch data
pluto.fetch_kaggle_dataset(
   'https://www.kaggle.com/datasets/tarkkaanko/amazon')
# import to Pandas
pluto.df_amazon_data = pluto.fetch_df(
   'kaggle/amazon/amazon_reviews.csv')
# count words and misspell
pluto.count_word(pluto.df_amazon_data,
   col_dest='reviewText')
pluto.check_spelling(pluto.df_amazon_data,
   col_dest='reviewText')
```

The data description of the Amazon reviews dataset on Kaggle is as follows:

"One of the most important problems in eCommerce is the correct calculation of the points given to after-sales products. The solution to this problem is to provide greater customer satisfaction for the eCommerce site, product prominence for sellers, and a seamless shopping experience for buyers. Another problem is the correct ordering of the comments given to the products. The prominence of misleading comments will cause both financial losses and customer losses."

The author is *Tarık kaan Koç*, and the license is **CC BY-NC-SA 4.0**: https://creativecommons.org/licenses/by-nc-sa/4.0/.

Pluto prints the batch using the following code:

```
# display misspelled words
pluto.print_batch_text(pluto.df_amazon_data,
   cols=['reviewText','misspelled'])
```

The result is as follows:

	reviewText	misspelled
4887	I got this for my Samsung Galaxy S5 and could not be happier. It's fast enough with the still and video camera on the phone.I have had no problems transferring files of all types from my pc or tablets. At this price it is a great buy.	{'s5', 'pc', 'samsung', 'phonei'}
4732	Great speed on both my DSLR and Smartphones. Great Sandisk quality as usual.Although other brands may be cheaper at times, I normally look to Sandisk first, and will usually buy it over other brands unless there is a large price difference.	{'dslr', 'sandisk', 'usualalthough'}
2385	im so upset right now it just died , turned on my phone and all my movies pics music all gone. card just died after three months bullcrap.	set()
2724	Works perfectly in my gopro black edition and offers a lot of space, very happy overall. would recommend to friends.	{'gopro'}
3393	Never had any issues with sandisk adapter card or devices. They work great and this one is no exception. Very small media for my big fingers but I manage. to do what I need to with it and it works well. Last thing is that Amazon's prices are hard to beat!	{'sandisk'}
985	If you carry two of these, one in your camera and one in the camera case pocket (don't lose it), you will never run out of memory. Enough capacity at ~4MB/image to shoot and shoot and shoot. Download at the end of the day if you want, but you have thousands of pictures on one of these. Couple with batteries (see WASABI review) and you can just never stop shooting.	{'4mbimage'}

Figure 2.17 – Amazon reviews misspelled words

Pluto has chosen to display two data columns in the print_batch function, as shown in *Figure 2.17*, but there are 12 data columns in the dataset. They are as follows:

- reviewerName
- overall
- reviewText
- reviewTime
- day_diff
- helpful_yes
- helpful_no
- total_vote
- score_pos_neg_diff
- score_average_rating
- wilson_lower_bound

Pluto draws the word counts and the misspelled words using the following code:

```
# display word count
pluto.draw_word_count(pluto.df_amazon_data)
# draw misspelled words
pluto.draw_word_count(pluto.df_amazon_data,
  wc='misspelled_count')
```

The result for the word counts is as follows:

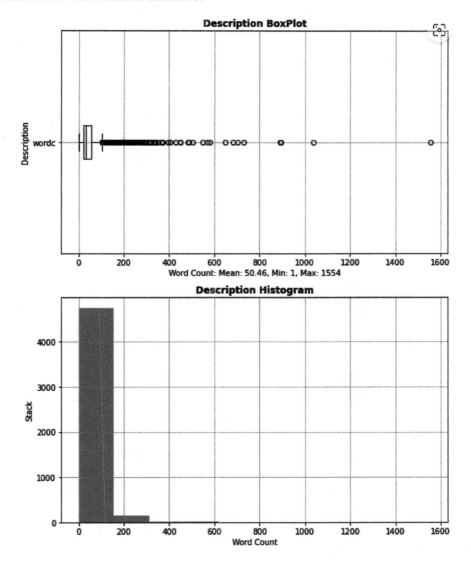

Figure 2.18 – Amazon reviews word count

Here is the graph for the misspelled words:

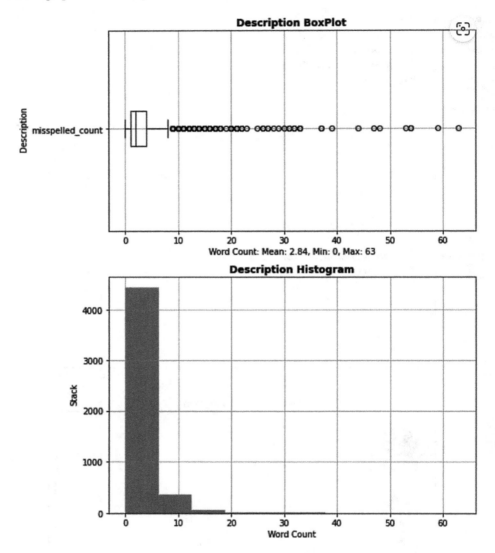

Figure 2.19 – Amazon reviews misspelled words graph

Pluto notices that the biases in the Amazon reviews, as shown in *Figures 2.17*, *2.18*, and *2.19*, are as follows:

- There are more grammatical errors in the Amazon reviews than in the Netflix movie description. Thus, there could be bias against well-written reviews.

- There are many more technical product names and jargon in the reviews. Therefore, there could be bias against non-technical reviewers.

- There are many outlines. The mean is 50.46 words per review, with the bulk feedback between 20 and 180 words. It is worth digging deeper using other columns, such as `helpful_yes`, `total_vote`, and `score_pos_neg_diff`, to see if there is bias in the review length per category.

- The Amazon reviews have more misspelled words than the Netflix movie description, reinforcing the well-written reviewer's bias.

Before jumping into the summary, here is a fun fact.

Fun fact

Cathy O'Neil's book, *Weapons of Math Destruction: How Big Data Increases Inequality and Threatens Democracy*, published in 2016, describes many biases in algorithms and AI, and it is a must-read for data scientists and college students. The two prominent examples are an accomplished teacher fired by a computer algorithm and a qualified college student rejected by the candidate screening software.

Summary

This chapter was not a typical one in this book because we discussed more theory than practical data augmentation techniques. At first, the link between data biases and data augmentation seems tenuous. Still, as you begin to learn about computational, human, and systemic biases, you see the strong connection because they all share the same goal of ensuring successful ethical AI system usage and acceptance.

In other words, data augmentation increases the AI's prediction accuracy while reducing the data biases in augmenting, ensuring the AI forecast has fewer false-negative and true-negative outcomes.

The computational, human, and systemic biases are similar but are not mutually exclusive. However, providing plenty of examples of real-world biases and observing three real-world image datasets and two real-world text datasets made these biases easier to understand.

The nature of data bias in augmenting makes it challenging to compute biases programmatically. However, you learned to write Python code for the fairness matrix in the text dataset using word counts and misspelled word techniques. You could use generative AI, such as Stable Diffusion or DALL-E, to automatically spot the biases in the photo and use OpenAI GPT3, GPT4, or Google Bard to compute the biases in text data. Unfortunately, generative AI is outside the scope of this book.

Initially, Pluto tended to go slow with step-by-step explanations, but as you learned, he shortened the justification and showed only the bare minimum code. The complete code can be found in the Python Notebook.

Most of the Python code is devoted to teaching you how to download real-world datasets from the *Kaggle* website and importing the metadata into pandas. The later chapters will reuse these helper and wrapper functions.

Throughout this chapter, there were *fun facts* and *fun challenges*. Pluto hopes you will take advantage of these and expand your experience beyond the scope of this chapter.

Pluto looks forward to *Chapter 3*, where he will play with image augmentation in Python.

Part 2: Image Augmentation

This part includes the following chapters:

3

Image Augmentation for Classification

Image augmentation in **machine learning** (**ML**) is a stable diet for increasing prediction accuracy, especially for the image classification domain. The causality logic is linear, meaning the more robust the data input, the higher the forecast accuracy.

Deep learning (**DL**) is a subset of ML that uses artificial neural networks to learn patterns and forecast based on the input data. Unlike traditional ML algorithms, which depend on programmer coding and rules to analyze data, DL algorithms automatically learn, solve, and categorize the relationship between data and labels. Thus, expanding the datasets directly impacts DL predictions on new insights that the model has not seen in the training data.

DL algorithms are designed to mimic the human brain, with layers of neurons that process information and pass it on to the next layer. Each layer of neurons learns to extract increasingly complex features from the input data, allowing the network to identify patterns and make predictions with increasing accuracy.

DL for image classification has proven highly effective in various industries, ranging from healthcare, finance, transportation, and consumer products to social media. Some examples include *identifying 120 dog breeds, detecting cervical spine fractures, cataloging landmarks, classifying Nike shoes, spotting celebrity faces*, and *separating paper and plastic for recycling*.

There is no standard formula to estimate how many images you need to achieve a designer prediction accuracy for image classification. Acquiring additional photos may not be a viable option because of cost and time. On the other hand, image data augmentation is a cost-effective technique that increases the number of photos for image classification training.

This chapter consists of two parts. First, you will learn the concepts and techniques of augmentation for image classification, followed by hands-on Python coding and a detailed explanation of the image augmentation techniques.

> **Fun fact**
>
> The image dataset is typically broken into 75% training, 20% validation, and 5% testing in the image classification model. Typically, the images allotted for training are augmented but outside the validation and testing set.

The two primary approaches for image augmentation are pre-processing and dynamic. They share the same techniques but differ when augmentation is done. The pre-processing method creates and saves the augmented photos in disk storage before training, while the dynamic method expands the input images during the training cycle.

In *Chapter 2*, you learned about data biases, and it is worth remembering that image augmentation will increase the DL model's accuracy and may also increase the biases.

In addition to biases, the other noteworthy concept is *safety*. It refers to the distortion magnitude that does not alter the original image label post-transformation. Different photo domains have different *safety* levels. For example, horizontally flipping a person's portrait photo is an acceptable augmentation technique, but reversing the hand gesture images in sign language is unsafe.

By the end of this chapter, you will have learned the concepts and hands-on techniques in Python coding for classification image augmentation using real-world datasets. In addition, you will have examined several Python open source libraries for image augmentation. In particular, this chapter covers the following topics:

- Geometric transformations
- Photometric transformations
- Random erasing
- Combining
- Reinforcing your learning through Python code

Geometric transformations are the primary image augmentation technique used commonly across multiple image datasets. Thus, this is a good place to begin discussing image augmentation.

Geometric transformations

Geometric transformation alters the photo's geometry, which is done by flipping along the X-axis or Y-axis, cropping, padding, rotating, warping, and translation. Complex augmentation uses these base photo-altering techniques. While working with geometric transformations, the distortion magnitude has to be kept to a safe level, depending on the image topic. Thus, no general formula governing geometric transformation applies to all photos. In the second half of this chapter, the Python coding section, you and Pluto will download real-world image datasets to define the safe level for each image set.

The following techniques are not mutually exclusive. You can combine horizontal flipping with cropping, resizing, padding, rotating, or any combination thereof. The one constraint is the safe level for distortion.

In particular, you will learn the following techniques:

- Flipping
- Cropping
- Resizing
- Padding
- Rotating
- Translation
- Noise injection

Let's start with flipping.

Flipping

The two flipping types are the horizontal Y-axis and the vertical X-axis. Turning the photos along the Y-axis is like looking in a mirror. Therefore, it can be used for most types of pictures except for directional images such as street signs. There are many cases where rotating along the X-axis is not safe, such as landscape or cityscape images where the sky should be at the top of the picture.

It is not an either-or proposition, and the image can use horizontal and vertical flips, such as aerial photos from a plane. Therefore, flipping is generally safe to use. However, some pictures are not safe for either transformation, such as street signs, where any rotation changes the integrity of the original label post-translation.

Later in this chapter, you will learn how to flip images using Python code with an image augmentation library, but for now, here is a teaser demonstration. The function's name is `pluto.draw_image_teaser_flip()`; the explanation will come later.

The image output is as follows:

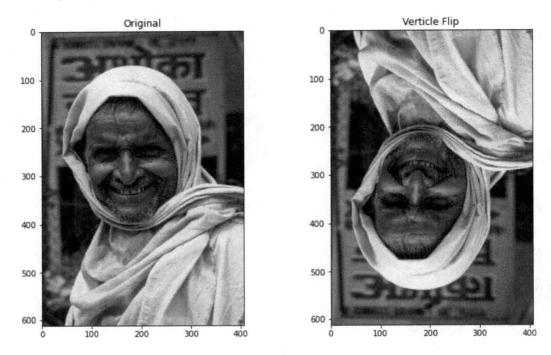

Figure 3.1 – Image vertical flip

Flipping keeps all the image content intact. However, the following technique, known as cropping, loses information.

Cropping

Cropping an image involves removing the edges of the picture. Most **convolutional neural networks** (**CNNs**) use a square image as input. Therefore, photos in portrait or landscape mode are regularly chopped to a square image. In most cases, the removal of the edges is based on the center of the picture, but there is no rule implying it has to be the center of the image.

The photo's center point is 50% of the width and 50% of the height. However, in image augmentation, you can choose to move the cropping center to 45% of the width and 60% of the height. The cropping center can vary, depending on the photo's subject. Once you have identified the safe range for moving the cropping center, you can try dynamically cropping the images per training epoch. Thus, every training epoch has a different set of photos. The effect is that the ML model will likely not overfit and gives higher accuracy from having more images.

The pluto.draw_image_teaser_crop() function is another teaser demonstration. Moving forward, I will only display the teaser images for some augmentation methods since you will learn about all of them in more depth by using Python code later in this chapter.

The output image for **center cropping** is as follows:

Figure 3.2 – Image center crop

Cropping is not the same as resizing an image, which we will discuss next.

Resizing

Resizing can be done by keeping the aspect ratio the same or not:

- **Zooming** is the same as enlarging, cropping, and maintaining the same aspect ratio.
- **Squishing** is the same as enlarging or shrinking and changing the original aspect ratio. The safe level for zooming, squishing, or other resizing techniques depends on the image category.

The `pluto.draw_image_teaser_resize()` function is a fun demonstration of **resizing** an image using the **squishing** mode. The output is as follows:

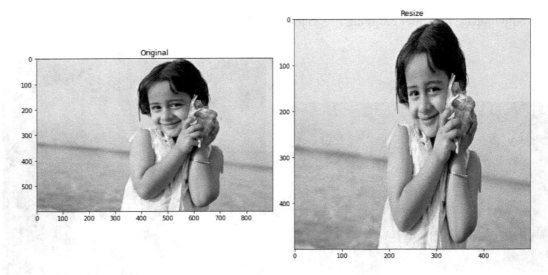

Figure 3.3 – Image resizing with squishing mode

When resizing a photo and not keeping the original aspect ratio, you need to pad the new image. There are different methods for **padding**.

Padding

Padding involves filling the outer edge of the canvas that is not an image. There are three popular methods for padding:

- **Zero padding** refers to padding the image with black, white, gray, or Gaussian noise
- **Reflection padding** mirrors the padding area with the original image
- **Border padding** involves repeating the borderline in the padding section

Padding is used in combination with cropping, resizing, translation, and rotating. Therefore, the safe proportion depends on cropping, resizing, and rotating.

Rotating

Rotating an image involves turning the picture clockwise or counterclockwise. The measurement of turning is by a degree and clockwise direction. Therefore, turning 180 degrees is the same as flipping vertically, while rotating 360 degrees returns the photo to its original position.

General rotating operates on the X-Y plane, whereas turning in the Z plane is known as **tilting**. **Skewing** or **shearing** involves rotating on all three planes – that is, X, Y, and Z. As with most geometric transformations, rotating is a safe operation with a set limit for some image datasets and not for others.

The `pluto.draw_image_teaser_rotate()` function is a fun demonstration of **rotating** an image with **reflection padding** mode. The output is as follows:

Figure 3.4 – Image rotating and reflection padding mode

Similar to rotating is shifting the images, which leads to the next technique, known as translation.

Translation

The translation method shifts the image left or right along the X-axis or up or down along the Y-axis. It uses padding to backfill the negative space left by shifting the photo. Translation is beneficial for reducing center image biases, such as when people's portraits are centered in the picture. The photo's subject will dictate the safe parameters for how much to move the images.

The next geometric transformation is different from the ones we've talked about so far because noise injection reduces the photo's clarity.

Noise injection

Noise injection adds random black, white, or color pixels to a picture. It creates a grainy effect on the original image. Gaussian noise is a de facto standard for generating natural noises in a photo. It is based on the Gaussian distribution algorithm developed by mathematician Carl Friedrich Gauss in the 1830s.

The `pluto.draw_image_teaser_noise()` function is a fun demonstration of **noise injection** using **Gaussian** mode. The output is as follows:

Figure 3.5 – Image noise injection using Gaussian mode

> **Fun challenge**
>
> Here is a thought experiment: can you think of other geometric image transformations? Hint: use the Z-axis, not just the X-axis and the Y-axis.

In the second part of this chapter, Pluto and you will discover how to code geometric transformations, such as flipping, cropping, resizing, padding, rotation, and noise injection, but there are a few more image augmentations techniques to learn first. The next category is photometric transformations.

Photometric transformations

Photometric transformations are also known as lighting transformations.

An image is represented in a three-dimensional array or a rank 3 tensor, and the first two dimensions are the picture's width and height coordinates for each pixel position. The third dimension is a **red, blue, and green** (RGB) value ranging from zero to 255 or #0 to #FF in hexadecimal. The equivalent of RGB in printing is **cyan, magenta, yellow, and key** (CMYK). The other popular format is **hue, saturation, and value** (HSV). The salient point is that a photo is a matrix of an integer or float when normalized.

Visualizing the image as a matrix of numbers makes it easy to transform it. For example, in HSV format, changing the **saturation** value to zero in the matrix will convert an image from color into grayscale.

Dozens of filters alter the color space characteristics, from the basics to exotic ones. The basic methods are **darkened**, **lightened**, **sharpened**, **blurring**, **contrast**, and **color casting**. Aside from the basics, there are too many filter categories to list here, such as **retro**, **groovy**, **steampunk**, and many others. Furthermore, photo software, such as Adobe Photoshop, and online image editors create new image filters frequently.

In particular, this section will cover the following topics:

- Basic and classic
- Advanced and exotic

Let's begin with basic and classic.

Basic and classic

Photometric transformations in image augmentation are a proven technique for increasing AI model accuracy. Most scholarly papers, such as *A comprehensive survey of recent trends in deep learning for digital images augmentation*, by Nour Eldeen Khalifa, Mohamed Loey, and Seyedali Mirjalili, published by *Artificial Intelligence Review* on September 4, 2021, use the classic filters exclusively because code execution is fast. There are many open source Python libraries for the classic filters, which Pluto and you will explore later in this chapter.

In particular, you will learn the following classic techniques:

- Darken and lighten
- Color saturation
- Hue shifting
- Color casting
- Contrast

Let's begin with the most common technique: the darken and lighten filter.

Darken and lighten

Lightening an image means increasing the brightness level, while lowering the brightness value means darkening an image. In Python code, a photo is an integer or float values matrix, and once converted into HSV format, raising or lowering the **value (V)** in the HSV matrix increases or decreases the picture's brightness level.

When it is time for you to write the functions for lightening or darkening the image for the Pluto object, you will use a Python image library to do the heavy lifting, but it is not hard to write the code from scratch. The safe range for the brightness value depends on the image subject and label target.

The pluto.draw_image_teaser_brightness() function is a fun demonstration of **darkening** an image. The output is as follows:

Figure 3.6 – Image brightness, darken mode

Similarly, color saturation is also easy to code in Python.

Color saturation

Color saturation involves increasing or decreasing the intensity of the color in a photo. By reducing the saturation values close to zero, the image becomes a grayscale image. Inversely, the picture will show a more intense or vibrant color when raising the saturation value.

Similar to the brightness level coding, manipulating the picture's **saturation** (**S**) value in the HSV matrix gives the desired effects. The safe range for color saturation depends on the image subject and label target.

So far, we've looked at the *S* and the *V* in *HSV*, but what does the *H* value do? It is for hue shifting.

Hue shifting

Shifting the **hue** (**H**) value in the Python image matrix in HSV format alternates the photo's color. Typically, a circle represents the hue values. Thus, the value starts at zero and ends at 360 degrees. Red is at the top of the rotation, beginning with zero, followed by yellow, green, cyan, blue, and magenta. Each color is separated by 60 degrees. Therefore, the last color, magenta, starts at 310 and ends at 360 degrees.

Hue shifting is an excellent image editing filter, but for AI image augmentation, it is not helpful because it distorts the image beyond the intended label. For example, suppose you are developing an AI model to classify different species of chameleons. In that case, the hue-switching technique is sufficient for image augmentation. Still, if your project is to differentiate cats and fluffy furball toys, it might lead to false positives because you would get fluorescent pink cats.

The `pluto.draw_image_teaser_hue()` function is a fun demonstration of **hue shifting**. The output is as follows:

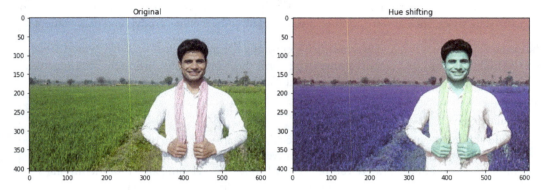

Figure 3.7 – Image hue shifting

Similar to hue shifting is color casting.

Color casting

Color casting is also known as color tinting. It is when the white color is not balanced or is inaccurate. The tint colors are commonly **red, green, or blue (RGB)**. In Python, tinting a photo is as easy as altering the RGB value in the image matrix. There is the same concern for the safe range as in the hue-shifting filter. In other words, color casting has limited use in AI image augmentation.

There is no formal definition of which filters are basic or classic and which are advanced and exotic. Hence, we have chosen to look at the contrast filter for our final example of classic photometric transformations.

Contrast

Contrast is the difference in luminance or color that distinguishes objects in a picture from one another. For example, most photographers want a high contrast between a person in the foreground concerning the background. Usually, the foreground object is brighter and in sharper focus than the background. The safe range for the contrast value depends on the image subject and label target.

Pluto and you will explore the contrast filter and all other classic photometric transformations using Python code in the second half of this chapter. The following section will cover advanced and exotic filters.

Advanced and exotic

The advanced or exotic techniques have no Python library for implementing these filters in data augmentation. Online photo editing websites and desktop software frequently create new exotic filters monthly.

If you review the filters section of *Adobe Photoshop* or many online photo editing websites, such as www.fotor.com, you will find dozens or hundreds of filter options. Specialized filters for image subjects include people portraits, landscapes, cityscapes, still life, and many others. Filters are also categorized by styles, such as retro, vintage, steampunk, trendy, mellow, groovy, and many others.

The exotic filters are not featured in scholarly papers partly due to the lack of available Python libraries and the high CPU or GPU resource time to perform these operations during the training cycle. Nevertheless, in theory, exotic filters are excellent techniques for image augmentation.

> **Fun challenge**
>
> Let's do a thought experiment. Can generative AI, like stable diffusion or DALL-E, create new images for augmentation? Generative AI can create hundreds or thousands of images from input text. For example, let's say you've been tasked with developing an AI for identifying a unicorn, pegasus, or minotaur; is it more difficult to find images of those mythical creatures in print or real life? Generative AI can do this, but is it a practical technique? Hint: think about static versus dynamic augmentation disk space and time.

Photometric and geometric transformations manipulate photos, but random erasing adds new elements to a picture.

Random erasing

Random erasing selects a rectangle region in an image and replaces or overlays it with a gray, black, white, or Gaussian noise pixels rectangle. It is counterintuitive to why this technique increases the AI model's forecasting accuracy.

The strength of any ML model, especially CNN, is in predicting or forecasting data that has not been seen in the training or validating stage. Thus, dropout, where randomly selected neurons are ignored during training, is a well-proven method to reduce overfitting and increase accuracy. Therefore, random erasing has the same effect as increasing the dropout rate.

A paper called *Random Erasing Data Augmentation*, which was published on November 16, 2017, by arXiv, shows how random erasing increases accuracy and reduces overfitting in a CNN-based model. The paper's authors are Zhun Zhong, Liang Zheng, Guoliang Kang, Shaozi Li, and Yi Yang from the Cognitive Science Department, at Xiamen University, China, and the University of Technology Sydney, Australia.

Typically, the random erasing rectangle region, also known as the cutout, is filled with random pixels using *Gaussian randomization*. The safe range for random erasing depends on the image subject and label target.

> **Fun fact**
>
> There is one creative example of using random erasing in image augmentation that reduces biases. In a self-driving automobile system, one of the image classification models is to identify and classify street signs. The AI model was trained with clear and pristine street sign photos, so the AI model was biased against the real-world pictures of street signs in poor neighborhoods in the USA, where street signs are defaced with graffiti and abused. Randomly adding cutouts of graffiti, paint, dirt, and bullet holes increased the model's accuracy and reduced overfitting and biases against poor neighborhood street signs.

Depending on the image dataset subject, random erasing, photometric, and geometric transformations can be mixed and matched. Let's discuss this in detail.

Combining

The techniques or filters in geometric transformations can be readily combined with most image topics. For example, you can mix horizontal flip, cropping, resizing, padding, rotation, translation, and noise injection for many domains, such as people, landscapes, cityscapes, and others.

In addition, taking landscape as a topic, you can combine many filters in photometric transformations, such as darkening, lightening, color saturation, and contrast. Hue shifting and color casting may not apply to landscape photos. However, advanced photographic transformation filters, such as adding rain or snow to landscape images, are acceptable.

There's more: you can add random erasing to landscape images. As a result, 1,000 landscape images may increase to 200,000 photos for training. That is the power of image augmentation.

> **Fun challenge**
>
> Here is a thought experiment: should you augment the entire image dataset or only a segment?
>
> Data augmentation can generate hundreds of thousands of new images for training, increasing AI prediction accuracy by decreasing the overfitting problem. But what if you also augmented the validation and testing dataset? Hint: think about real-world applications, DL generalization, and false negatives and false positives.

So far, we have discussed various image augmentation filters and techniques. The next step is for you and Pluto to write Python code to reinforce your understanding of these concepts.

Reinforcing your learning through Python code

We will pursue the same approach as in *Chapter 2*. Start by loading the data_augmentation_
with_python_chapter_3.ipynb file in Google Colab or your chosen Jupyter Notebook or
JupyterLab environment. From this point onward, the code snippets will be from the Python Notebook,
which contains all the functions.

This chapter's coding lessons topics are as follows:

- Pluto and the Python Notebook
- Real-world image dataset
- Image augmentation library
- Geometric transformations
- Photometric transformations
- Random erasing
- Combining

The next step is to download, set up, and verify that Pluto and the Python Notebook are working adequately.

Pluto and the Python Notebook

Before loading Pluto from *Chapter 2*, we must retrieve him by cloning this book's **GitHub** repository.
Using the Python Notebook's %run magic command, we can invoke Pluto. If you improved or hacked
Pluto, load that file. You should review *Chapter 2* if these steps are not familiar to you.

> **Fun fact**
>
> The startup process for coding is the same for every chapter. Pluto only displays the essential code
> snippets in this book, and he relies on you to review the complete code in the Python Notebook.

Use the following code to clone the Python Notebook and invoke Pluto:

```
# clone the book repo.
f = 'https://github.com/PacktPublishing/Data-Augmentation-with-Python'
!git clone {f}
# invoke Pluto
%run 'Data-Augmentation-with-Python/pluto/pluto_chapter_2.py'
```

The output will be similar to the following:

```
------------------------------ : ---------------------------
            Hello from class : <class '__main__.PacktDataAug'> Class:
PacktDataAug
                 Code name : Pluto
                 Author is : Duc Haba
------------------------------ : ---------------------------
```

Double-check that Pluto has loaded correctly by running the following code in the Python Notebook:

```
# display system and libraries version
pluto.say_sys_info()
```

The output will be as follows or something similar, depending on your system:

```
------------------------------ : ---------------------------
                  System time : 2022/09/18 06:07
                     Platform : linux
      Pluto Version (Chapter) : 2.0
        Python version (3.7+) : 3.7.13 (default, Apr 24 2022, 01:04:09)
[GCC 7.5.0]
             PyTorch (1.11.0) : actual: 1.12.1+cu113
              Pandas (1.3.5) : actual: 1.3.5
                 PIL (9.0.0) : actual: 7.1.2
          Matplotlib (3.2.2) : actual: 3.2.2
                   CPU count : 2
                   CPU speed : NOT available
------------------------------ : ---------------------------
```

Pluto has verified that the Python Notebook is working correctly, so the next step is downloading real-world image datasets from the *Kaggle* website.

Real-world image datasets

In *Chapter 2*, Pluto learned how to download thousands of real-world datasets from the *Kaggle* website. For this chapter, he has selected six image datasets to illustrate different image augmentation techniques. Still, you can substitute or add new *Kaggle* image datasets by passing the new URLs to the code in the Python Notebook.

> **Fun challenge**
>
> Download two additional real-world datasets from the Kaggle website. Pluto likes to play fetch, so it is no problem for it to fetch new datasets. Hint: go to https://www.kaggle.com/datasets and search for **image classification**. Downloading additional real-world data will further reinforce your understanding of image augmentation concepts.

Pluto has chosen six image datasets based on the challenges each topic brings to bear on augmentation techniques. In other words, one concept may be acceptable for one subject but not for another. In particular, the six image datasets are as follows:

- Covid-19 image dataset

- Indian people

- Edible and poisonous fungi

- Sea animals

- Vietnamese food

- Mall crowd

> **Fun fact**
>
> The code for downloading these six real-world datasets from the *Kaggle* website looks repetitive. It is easy by design because Pluto worked hard to create reusable methods in *Chapter 2*. He wants it to be easy so that you can download any real-world dataset from the *Kaggle* website.

Let's start with the Covid-19 data.

Covid-19 image dataset

Medical is a popular category for AI image predictive models. Therefore, Pluto selected the *Covid-19 Image Dataset*. He fetched the pictures and made the necessary pandas DataFrame using the methods shown in *Chapter 2*. Note that the complete code is in the Python Notebook.

The following commands fetch and load the data into pandas:

```
# fetch image data
pluto.fetch_kaggle_dataset('https://www.kaggle.com/datasets/
pranavraikokte/covid19-image-dataset')
# import to Pandas data frame
f = 'kaggle/covid19-image-dataset/Covid19-dataset/train'
pluto.df_covid19 = pluto.make_dir_dataframe(f)
```

The first three records of the pandas DataFrame are as follows:

	fname	label
0	kaggle/covid19-image-dataset/Covid19-dataset/t...	Covid
1	kaggle/covid19-image-dataset/Covid19-dataset/t...	Covid
2	kaggle/covid19-image-dataset/Covid19-dataset/t...	Covid

Figure 3.8 – The first three rows of the pandas DataFrame

On the *Kaggle* website, the data's context is as follows:

> *"Helping Deep Learning and AI Enthusiasts like me to contribute to improving
> Covid-19 detection using just Chest X-rays. It contains around 137 cleaned images
> of Covid-19 and 317 containing Viral Pneumonia and Normal Chest X-Rays
> structured into the test and train directories."*

This citation is from the *University of Montreal*, and the collaborator listed is **Pranav Raikote** (owner), license: **CC BY-SA 4.0**: https://choosealicense.com/licenses/cc-by-sa-4.0.

Now that Pluto has downloaded the Covid-19 data, it will start working on the *People* dataset.

Indian People

The second typical category in image prediction or classification is people. Pluto has chosen the *Indian People* dataset. The following code snippet from the Python Notebook fetches and loads the data into pandas:

```
# fetch image data
pluto.fetch_kaggle_dataset('https://www.kaggle.com/datasets/
sinhayush29/indian-people')
# import to Pandas DataFrame
f = 'kaggle/indian-people/Indian_Train_Set'
pluto.df_people = pluto.make_dir_dataframe(f)
```

On the *Kaggle* website, there is no description of the dataset. It's not uncommon to get a dataset without an explanation or goals and be asked to augment it. The collaborator listed is **Ayush Sinha** (owner), license: **None, Visible to the public**.

The typical usage for people data is to identify or classify age, sex, ethnicity, emotional sentiment, facial recognition, and many more.

Fun fact

There are controversial image classification AI systems, such as those for predicting people as criminals or not criminals, forecasting worthiness to society, identifying sexual orientation, and selecting immigrants or citizens. However, other creative uses include identifying a potential new whale – a super high casino spender from a casino installing cameras in the lobby and feeding them to an AI.

Now, let's look at the fungi data.

Edible and poisonous fungi

The third most often used topic for image classification is safety, such as distracted drivers, poisonous snakes, or cancerous tumors. Pluto found the real-word *Edible and Poisonous Fungi* dataset on the *Kaggle* website. The following code snippet from the Python Notebook fetches and loads the data into pandas:

```
# fetch image data
pluto.fetch_kaggle_dataset('https://www.kaggle.com/datasets/
marcosvolpato/edible-and-poisonous-fungi')
# import into Pandas data frame
f = 'kaggle/edible-and-poisonous-fungi'
pluto.df_fungi = pluto.make_dir_dataframe(f)
```

On the *Kaggle* website, the description is as follows:

> *"We created this dataset as part of our school's research project. As we didn't find something similar when we started, we decided to publish it here so that future research with mushrooms and AI can benefit from it."*

The collaborator listed is **Marcos Volpato** (owner), license: **Open Data Commons Open Database License (ODbL)**: https://opendatacommons.org/licenses/odbl/1-0/.

Now, let's look at the sea animals data.

Sea animals

The fourth theme is nature. Pluto selected the *Sea Animals Image Dataset*. The following commands fetch and load the data into pandas:

```
# fetch image data
pluto.fetch_kaggle_dataset('https://www.kaggle.com/datasets/
vencerlanz09/sea-animals-image-dataste')
# import to Pandas data frame
f = 'kaggle/sea-animals-image-dataste'
pluto.df_sea_animal = pluto.make_dir_dataframe(f)
```

The *Kaggle* website's description for this dataset is as follows:

> *"Most life forms began their evolution in aquatic environments. The oceans provide about 90% of the world's living space in terms of volume. Fish, which are only found in water, are the first known vertebrates. Some of these transformed into amphibians, which dwell on land and water for parts of the day."*

The collaborators listed are **Vince Vence** (owner), license: **Other— Educational purposes and Free for Commercial Use (FFCU)**: https://www.flickr.com/people/free_for_commercial_use/.

Next, we'll look at food data.

Vietnamese food

The fifth widespread subject for image classification is food. Pluto found the *30VNFoods – A Dataset for Vietnamese Food Images Recognition* dataset. The following commands fetch and load the data into pandas:

```
# fetch image data
pluto.fetch_kaggle_dataset('https://www.kaggle.com/datasets/quandang/
vietnamese-foods')
# import to Pandas DataFrame
f = 'kaggle/vietnamese-foods/Images/Train'
pluto.df_food = pluto.make_dir_dataframe(f)
```

The *Kaggle* website's description is as follows:

> "This paper introduces a large dataset of 25,136 images of 30 popular Vietnamese foods. Several machine learning and deep learning image classification techniques have been applied to test the dataset, and the results were compared and reported."

The collaborators listed are **Quan Dang** (owner), **Anh Nguyen Duc Duy** (editor), **Hoang-Nhan Nguyen** (viewer), **Phuoc Pham Phu** (viewer), and **Tri Nguyen** (viewer), license: **CC BY-SA 4.0**: https://choosealicense.com/licenses/cc-by-sa-4.0.

Now, let's move on to mall crowd data.

Mall crowd

Pluto chose the sixth and last dataset for the creative use of AI image classification – the *Mall - Crowd Estimation* dataset. The following code snippet from the Python Notebook fetches and loads the data into pandas:

```
# fetch image data
pluto.fetch_kaggle_dataset('https://www.kaggle.com/datasets/
ferasoughali/mall-crowd-estimation')
# import to Pandas DataFrame
f = 'kaggle/mall-crowd-estimation/mall_dataset/frames'
pluto.df_crowd = pluto.make_dir_dataframe(f)
```

The *Kaggle* website's description is as follows:

> "The mall dataset was collected from a publicly accessible webcam for crowd counting and profiling research."

The collaborator listed is **Feras** (owner), license: **None, Visible to the public**.

Fun challenge

Refactor the code provided and write one function that downloads all six datasets. Hint: put the six *Kaggle* data URLs into an array. Pluto does not write the uber-big method because he focuses on making the augmentation techniques easier to understand rather than writing compact code that might obfuscate the meaning.

After downloading all six datasets, Pluto must draw an image batch.

Drawing an image batch

Let's look at the pictures in the six datasets. Pluto will take samples from the pandas DataFrame and use the `draw_batch()` function defined in *Chapter 2*.

The output for two Covid-19, two people, two fungi, two sea animals, one food, and one mall crowd picture are as follows:

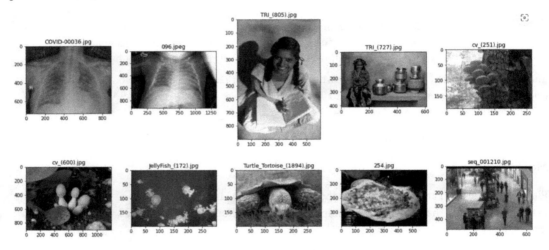

Figure 3.9 – Six image datasets

Pluto has downloaded plenty of real-world pictures, so the next step is selecting an image augmentation library.

Image augmentation library

There are many open source Python image augmentation and processing libraries. Most libraries have filters for geometric and photometric transformations. In addition, a few libraries have specialized functions for particular image topics.

Pluto will cover only some of these libraries. The most popular libraries are Albumentations, Fast.ai, **Pillow** (**PIL**), OpenCV, scikit-learn, Mahotas, and pgmagick:

- **Albumentations** is a fast and highly customizable image augmentation Python library. It has become the de facto standard for research areas related to computer vision and DL. Albumentations efficiently implements over 70 varieties of image transform operations optimized for performance. Albumentations' substantial benefit is broad integration with many DL frameworks. It was introduced in 2019. It can be found on GitHub at `https://github.com/albumentations-team/albumentations`.

- **Fast.ai** is a best-of-class for DL and AI library and framework. It was founded in 2016 by Jeremy Howard and Rachel Thomas to democratize DL. Fast.ai has extensive built-in functions for image augmentation. Furthermore, its image augmentation operations use GPU, so it is possible to perform dynamic image augmentation during the training cycle. In other words, because of the GPU, it is the best performance image augmentation library in the market. It can be found on GitHub at `https://github.com/fastai`.

- **Pillow** is a friendly modern fork of the **Python Imaging Library** (**PIL**) repository. PIL is a popular open source library for image processing and augmentation because it was first released in 1995. Many open source Python image processing, displaying, and augmenting libraries are built on top of PIL. It can be found on GitHub at `https://github.com/python-pillow/Pillow`.

- **AugLy** is an open source Python project by Meta (Facebook) for data augmentation. The library provides over 100 audio, video, image, and text data augmentation methods. It can be found on GitHub at `https://github.com/facebookresearch/AugLy`.

- **OpenCV** was developed by Intel in 2000 as an open source library. ML primarily uses OpenCV in computer vision tasks such as object classification and detection, face recognition, and image segmentation. In addition, OpenCV contains essential methods for ML. It can be found on GitHub at `https://github.com/opencv/opencv`.

- **scikit-learn** was one of the early open source libraries in 2009 for image augmentation. Part of scikit-learn is written in Cython, a programming language that is a superset of Python. One of its crucial benefits is high-performance speed, where a NumPy array is used as the image's structure. It can be found on GitHub at `https://github.com/scikit-image/scikit-image`.

- **Mahotas** is an image processing and augmentation library specialized in bioimage informatics. Mahotas uses NumPy arrays and is written C++ with a Python interface. It was released in 2016. It can be found on GitHub at `https://github.com/luispedro/mahotas`.

- **pgmagick**: pgmagick is a GraphicsMagick binding for Python. GraphicsMagick is best known for supporting large images in a gigapixel-size range. It was initially derived from ImageMagick in 2002. It can be found on GitHub at `https://github.com/hhatto/pgmagick`.

No library is better than another, and you can choose to use multiple libraries in a project. However, Pluto recommends picking two or three libraries and becoming proficient, maybe even an expert, in them.

Pluto will hide the library or libraries and create a wrapper function, such as `draw_image_flip()`, that uses other libraries to perform the transformation. The other reason for writing wrapper functions is to switch out the libraries and minimize the code changes. Pluto has chosen the **Albumentations**, **Fast.ai**, and **PIL** libraries for this chapter as the under-the-hood engine.

You have two options: creating image augmentation dynamically per batch or statically. When doing this statically, also known as pre-processing, you create and save the augmented pictures in your local or cloud drive.

For this chapter, Pluto has chosen to augment the image dynamically because, depending on the combinations of filters, you can generate over a million acceptable altered pictures. The only difference between the two methods is that the pre-processing method saves the augmented photos in local or cloud drives while the dynamic method does not.

> **Fun challenge**
>
> Here is a thought experiment: should you select an augmentation library with more augmented methods over a library that runs on GPU? Hint: think about the goal of your project and its disk and time resources.

Let's begin writing code for the geometric transformation filters.

Geometric transformation filters

Pluto can write Python code for many geometric transformation filters, and he will select two or three image datasets to illustrate each concept. In addition, by using multiple image subjects, he can discover the safe level. The range for the safe level is subjective, and you may need to consult a domain subject expert to know how far to distort the photo. When convenient, Pluto will write the same method using different libraries.

Let's get started with flipping images.

Flipping

Pluto will begin with the simplest filter: the horizontal flip. It mirrors the image, or in other words, it flips the photo along the Y-axis. The wrapper function is called `draw_image_flip()`. All image augmentation methods are prefixed with `draw_image_` as this makes it easy for Pluto to remember them. In addition, he can use the Python Notebook auto-complete typing feature. By typing `pluto.draw_im`, a popup menu containing all the filter functions will be displayed.

In the `draw_image_flip_pil()` function, when using the PIL library, the relevant code line is as follows:

```
# use PIL to flip horizontal
mirror_img = PIL.ImageOps.mirror(img)
```

Thus, Pluto selects an image from the People dataset and flips it using the following code:

```
# Select an image from Pandas and show the original and flip
pluto.draw_image_flip_pil(pluto.df_people.fname[100])
```

The result is as follows, with the original image at the top and the flip image at the bottom:

Figure 3.10 – Horizontal flip using the PIL library

Rather than viewing one image at a time, it is more advantageous to examine the entire dataset one batch at a time. This is because a filter may be applicable, or the **safe** range is acceptable for one image but not for another in the same dataset. The **Fast.ai** library has the data-batch class that supports many ML functions, including accepting a transformation method and displaying a random collection of pictures, also known as displaying a batch.

Pluto will write two new methods: `_make_data_loader()`, which is a helper function for creating the **Fast.ai** data-loader object, and the `draw_image_flip()` function, which encodes the transformation for horizontal flip and displays the image batch using the data-loader `show_batch()` method.

`show_batch()` will select a random set of pictures to display, where `max_n` sets the number of images in a bunch. The Fast.ai transformation, by default, performs the modification at 75% probability. In other words, three out of four images in the dataset will be transformed.

The horizontal flip filter has no safe level, regardless of whether it applies to the image set. Pluto will use the `draw_image_flip()` method with the **Fast.ai** transformation. The coding for all wrapper functions is very similar. Only the augmentation function, the `aug` value, is different. The entirety of the flip wrapper code is as follows:

```
# use fast.ai to flip image in wrapper function
def draw_image_flip(self,df,bsize=15):
  aug = [fastai.vision.augment.Flip(p=0.75)]
  item_fms = fastai.vision.augment.Resize(480)
  dsl_org = self._make_data_loader(df, aug,item_fms)
  dsl_org.show_batch(max_n=bsize)
  return dsl_org
```

The definition of the `aug` variable differs from one wrapper function to another. Pluto needs to run a function on the Python Notebook for the People dataset with the following code:

```
# Show random flip-image batch 15 in Pandas, wraper function
pluto.dls_people = pluto.draw_image_flip(pluto.df_people)
```

The result is as follows:

Figure 3.11 – Horizontal flip on the People dataset

> **Fun fact**
>
> The complete fully functional object-oriented code can be found in the Python Notebook. You can hack it to show flip, rotate, tilt, and dozens of other augmentation techniques.

To ensure the horizontal flip is acceptable, you can repeatedly run the `draw_image_flip()` function in the Python Notebook to see a collection of varying image batches. Horizontal flip is a safe filter for fungi, sea animals, food, and mall crowd pictures. Common sense dictates that you wouldn't expect otherwise. Here is the command for the Fungi dataset:

```
# use flip wrapper function on Fungi data
pluto.dls_fungi = pluto.draw_image_flip(pluto.df_fungi)
```

The result is as follows:

Figure 3.12 – Horizontal flip on the Fungi dataset

For medical images, such as the Covid-19 photos, you need a domain expert to confirm that flipping horizontally does not change the image's integrity. It does not make any difference to the layman, but it can be deceptively wrong and might create a **false-positive** or **false-negative** prediction. Here is the command for it:

```
# use flip wrapper function on covid data
pluto.dls_covid19 = pluto.draw_image_flip(pluto.df_covid19)
```

The result is as follows:

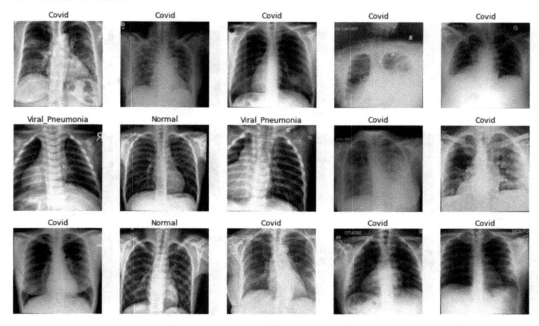

Figure 3.13 – Horizontal flip on the Covid-19 dataset

Notice that the Fast.ai transformation cropped the images center square. Therefore, in some pictures, content is lost – for example, a picture of a woman with most of her face missing. This is because Fast.ai is for ML, so the images need to be square. The default behavior is a center square crop.

Before Pluto can start cropping and padding, he must complete the flipping filter by combining horizontal flipping with vertical flipping. The people, Covid-19, fungi, and mall crowd pictures cannot be flipped vertically, but the sea animals and food pictures can.

For this, Pluto needs to create the draw_image_flip_both() method, with the transformation set to the following:

```
# using fast.ai for fliping
aug = fastai.vision.augment.Dihedral(p=0.8,pad_mode=pad_mode)
```

Now, Pluto must run the function on the People dataset with the following code:

```
# use wrapper function on both flip on people images
pluto.dls_people = pluto.draw_image_flip_both(
  pluto.df_people)
```

The result is as follows:

Figure 3.14 – Unsafe horizontal and vertical flips on the People dataset

He can apply the same function to the food pictures, as follows:

```
# use flip wrapper function on food photos
pluto.dls_food = pluto.draw_image_flip_both(pluto.df_food)
```

The result is as follows:

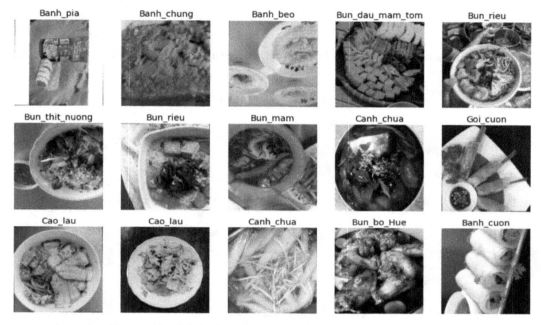

Figure 3.15 – Safe horizontal and vertical flips on the food dataset

Fun fact

Pluto loves to play the same game over and over again. You know, because he is a dog. :-) Thus, you can ask Pluto to run any wrapper functions repeatedly on the Python Notebook to see a different set of image batches from the data stored in pandas. Each real-world image dataset contains thousands of photos, and each batch displays 15 images; therefore, you must run the wrapper functions repeatedly to have a good mental picture of the dataset.

The next filters we will look at are for cropping and padding.

Cropping and padding

Reusing the same process as when writing the flipping filter, Pluto can write the draw_image_crop() method. The one new code line uses a different item transformation:

```
# use fast.ai to crop image in wrapper function
item_tfms=fastai.vision.augment.CropPad(480,
    pad_mode="zeros")
```

The padding mode can be **zeros**, which means the padding color is black, **border**, which means the padding repeats the border pixel, or **reflection**, which means padding is mirrored from the picture.

After much trial and error on the Python Notebook, Pluto found the safe range for cropping and padding for each of the six datasets. Before moving on, Pluto encourages you to use the Python Notebook to find the best safe parameter.

Pluto found that the safe setting for the people data is using a cropped image size of 640 and pad mode on the border:

```
# use wrapper function to crop and pad people photo
pluto.dls_people = pluto.draw_image_crop(pluto.df_people,
  pad_mode="border",
  isize=640)
```

The result is as follows:

Figure 3.16 – Horizontal and vertical flip on the People dataset

In terms of the next dataset, Pluto found that the safe setting for the fungi images is a cropped image size of 240 and a pad mode of zeros:

```
# use wrapper function to crop and pad Fungi image
pluto.dls_fungi = pluto.draw_image_crop(pluto.df_fungi,
   pad_mode="zeros",
   isize=240)
```

The result is as follows:

Figure 3.17 – Horizontal and vertical flip on the fungi dataset

For the food pictures, Pluto discovered that the safe parameters are a cropped image size of 640 and a pad mode of reflection:

```
# use wrapper function to crop and pad food image
pluto.dls_food = pluto.draw_image_crop(pluto.df_food,
   pad_mode="reflection",
   isize=640)
```

The result is as follows:

Figure 3.18 – Horizontal and vertical flip on the food dataset

> **Fun challenge**
>
> Find the safe cropping parameters for all six image datasets. You will get bonus points for applying these new functions to the set of images from your project or downloading them from the Kaggle website.

For the other image datasets, the results are in the Python Notebook. The safe parameter is 340 pixels with reflection padding for the sea animals pictures and 512 pixels with border padding for the mall crowd pictures. A cropping filter is not an option for the Covid-19 pictures.

Next, Pluto will rotate images, which is similar to flipping.

Rotating

Rotation specifies how many degrees to turn the image clockwise or counter-clockwise. Since Pluto sets the max rotation value, the actual rotation is a random number between the minimum and the maximum value. The minimum default value is zero. Therefore, a higher maximum value will generate more augmentation images because every time the system fetches a new data batch, a different rotation value is chosen. In addition, randomness is the reason for selecting the dynamic augmentation option over saving the images to a local disk drive, as in the static option.

For this, Pluto has written the `draw_image_rotate()` method using the `max_rotate = max_rotate` transformation parameter, where the second `max_rotate` is the passed-in value. The key code line in the wrapper function is as follows:

```
# use fast.ai for rotating
aug = [fastai.vision.augment.Rotate(max_rotate,
  p=0.75,
  pad_mode=pad_mode)]
```

Once again, Pluto has arrived at the following safe parameter for rotating after much trial and error on the Python Notebook, but don't take Pluto's word for it. Pluto challenges you to find better safe parameters by experimenting with the Python Notebook.

For the sea animals data, Pluto has arrived at a safe parameter of `180.0` for the maximum rotation and reflection for padding. The command in the Python Notebook is as follows:

```
# use wrapper function to rotate sea animal photo
pluto.dls_sea_animal = pluto.draw_image_rotate(
  pluto.df_sea_animal,
  max_rotate=180.0,
  pad_mode='reflection')
```

The result is as follows:

Figure 3.19 – Rotation on the sea animals dataset

For the people pictures, Pluto has arrived at a safe parameter of 25.0 for the maximum rotation and border for padding. The command in the Python Notebook is as follows:

```
# user wrapper function to rotate people photo
pluto.dls_people = pluto.draw_image_rotate(pluto.df_people,
  max_rotate=25.0,
  pad_mode='border')
```

The result is as follows:

Figure 3.20 – Rotation on the People dataset

For the other image datasets, the results are in the Notebook. The safe parameters are 16.0 maximum rotation with border padding for the mall crowd photos, 45.0 maximum rotation with border padding for the fungi pictures, 90.0 maximum rotation with reflection padding for the food images, and 12.0 maximum rotation with border zeros for the Covid-19 data. I encourage you to extend beyond Pluto's safe range on the Notebook and see what happens to the pictures for yourself.

Now, let's continue with the shifting image theme. The next filter we'll look at is the translation filter.

Translation

The translation filter shifts the image to the left, right, up, or down. It is not one of the commonly used filters. Pluto uses the `ImageChops.offset()` method in the **PIL** library to write the `draw_image_shift()` function. A negative horizontal shift value moves the image to the left, a positive value moves the image to the right, and the vertical shift parameter moves the image up or down. The relevant code line in the wrapper function is as follows:

```
# using PIL for shifting image
shift_img = PIL.ImageChops.offset(img,x_axis,y_axis)
```

To test the function, Pluto selects a picture and shifts it left by 150 pixels and up by 50 pixels. The command is as follows:

```
# select an image in Pandas
f = pluto.df_people.fname[30]
# user wrapper function to shift the image
pluto.draw_image_shift_pil(f, -150, -50)
```

The output is as follows:

Figure 3.21 – Translation using the PIL library; the top image is the original

The translation filter is seldom used because it is easy to find the safe level for one picture but not for the entire dataset.

So far, Pluto has shown you the **flipping**, **cropping**, **padding**, **resizing**, **rotating**, and **translation** filters. However, there are many more geometric transformation filters, such as for **warping**, **zooming**, **tilting**, and **scaling**. Unfortunately, there are too many to cover, but the coding process is the same.

> **Fun challenge**
> Implement two more geometric transformation techniques, such as warping and tilting. Hint: copy and paste from Pluto's wrapper functions and change the `aug` and `item_tfms` variables.

Moving from geometric to photometric transformations follows the same coding process. First, Pluto writes the wrapper functions using the Albumentations library, then uses the real-world image dataset to test them.

Photographic transformations

Pluto chose the Albumentations library to power the photometric transformations. The primary reasons are that the **Albumentations** library has over 70 filters, and you can integrate it into the **Fast.ai** framework. Fast.ai has most of the basic photometric filters, such as hue shifting, contrast, and lighting, but only Albumentations has more exotic filters, such as those for adding rain, motion blur, and FancyPCA. Be careful when using fancy filters. Even though they are easy to implement, you should research AI scholarly published papers to see if the filter is beneficial for achieving a higher accuracy rate.

As with the geometric transformations coding process, Pluto creates the base method and writes the wrapper function for each photometric transformation. The `_draw_image_album()` method is used to select a sample set of images from the data, convert it into a numpy array, do the transformation, and display them in batch mode. The pertinent code for the `_draw_image_album()` function is as follows:

```
# select random images
samp = df.sample(int(ncol * nrow))
# convert to an array
img_numpy = numpy.array(PIL.Image.open(samp.fname[i]))
# perform the transformation using albumentations
img = aug_album(image=img_numpy)['image']
# display the image in batch modde
ax.imshow(img)
```

The wrapper function code is straightforward. For example, the code for the brightness filter is as follows:

```python
def draw_image_brightness(self,df,
  brightness=0.2,
  bsize=5):
  aug_album = albumentations.ColorJitter(
    brightness = brightness,
    contrast=0.0,
    saturation=0.0,
    hue=0.0,
    always_apply=True,
    p=1.0)
  self._draw_image_album(df,aug_album,bsize)
  Return
```

> **Fun fact**
>
> For any of the Albumentations functions, you can append a question mark (?) and run the code cell to see the documentation in the Python Notebook; for example, `albumentations.ColorJitter?`. Append two question marks (??) to see the function's Python source code; for example, `albumentations.ColorJitter??`. A bonus fun fact is that Albumentations types are followed by a dot – for example, `albumentations.` – in the Python Notebook and wait for a second. A list of all the available functions appears in a drop-down list, where you can choose one. In other words, the Python Notebook has auto-complete typing.

The definition of the `aug_albm` variable differs from one wrapper function to another. Let's test out the brightness filter.

Brightness

It isn't easy to view most of the photometric transformations in grayscale because you are reading them in a book. That is more reason for joining Pluto with coding in the Notebook as you can see color. Rather than showing you all optimal safe parameters for each filter, Pluto will show you the *unsafe* range for one dataset and a safe parameter for the other dataset. The key code line for brightness in the wrapper function is as follows:

```python
# use the Albumentations library function for brightness
aug_album = albumentations.ColorJitter(brightness=brightness,
  contrast=0.0,
  saturation=0.0,
  hue=0.0,
  always_apply=True,
  p=1.0)
```

You can see the mistake in this book by exaggerating the brightness – for example, if it's too bright or too dark. Once again, you should look at the Notebook to see the brightness effects in color. For the people photos, the *unsafe* value is a brightness equal to 1.7. Pluto runs the following command on the Python Notebook:

```
# use the brightness wrapper function for people photo
pluto.draw_image_brightness(pluto.df_people, brightness=1.7)
```

The output is as follows:

Figure 3.22 – Unsafe brightness level for the People dataset

The People dataset does not have any objective. Therefore, it is challenging to find safe parameters. If the goal is as simple as classifying people's ages, the brightness level can be relatively high, but without knowing the intended use of the dataset, you don't know how much to distort the pictures.

Pluto found that the safe brightness value for the food dataset is 0.3, but it may not be easy to see the effects in this book. Here is the command that he used on the Python Notebook:

```
# use the brightness wrapper function for food image
pluto.draw_image_brightness(pluto.df_food, brightness=0.3)
```

The output is as follows:

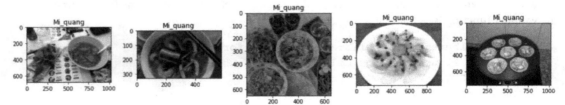

Figure 3.23 – Safe brightness level for the food dataset

The brightness level for the other four datasets is similar. Pluto has left it up to you to experiment and find the safe level in the Python Notebook. The Covid-19 images are in grayscale, and the intent is to predict Covid-19 patients from their chest X-ray photos. A decrease or increase in the brightness level may result in a false-positive or false-negative prediction. You should consult with the domain experts to confirm the safe parameters for Covid-19 images.

The grayscale filter wasn't discussed in the first half of this chapter, but it is similar to the brightness filter.

Grayscale

A few scholarly papers describe the benefit of grayscale, such as *Data Augmentation Methods Applying Grayscale Images for Convolutional Neural Networks in Machine Vision*, by Jinyeong Wang and Sanghwan Lee in 2021 from the Department of *Mechanical Convergence Engineering, Hanyang University, Seoul 04763, Korea*. The paper explains the effective data augmentation method for grayscale images in CNN-based machine vision with mono cameras.

In the `draw_image_grayscale()` method, Pluto uses the Albumentations library function as follows:

```
# use albumentations for grayscale
aug_album = albumentations.ToGray(p=1.0)
```

The fungi dataset aims to classify whether a mushroom is edible or poisonous, and the mushroom's color significantly affects the classification. Therefore, converting into grayscale is not advisable. Nevertheless, Pluto illustrates the grayscale filter with the following command:

```
# use the grayscale wrapper function for fungi image
pluto.draw_image_grayscale(pluto.df_fungi)
```

The output is as follows:

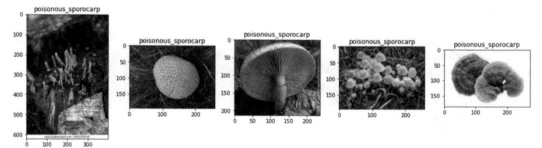

Figure 3.24 – Unsafe use of grayscale on the fungi dataset

The mall crowd dataset's goal is to estimate the crowd size in a shopping mall. Thus, converting the photos into grayscale should not affect the prediction. Pluto runs the following command:

```
# use the grayscale wrapper function for crowd photo
pluto.draw_image_grayscale(pluto.df_crowd)
```

The results are as follows:

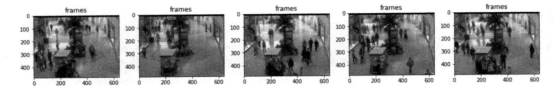

Figure 3.25 – Safe use of grayscale on the mall crowd dataset

Pluto left the other four datasets for you to experiment with to determine whether it is safe to use the grayscale filter. After you use the Python Notebook to explore these datasets, come back here, where we will examine the contrast, saturation, and hue-shifting filters.

Contrast, saturation, and hue shifting

The contrast, saturation, and hue-shifting filters are beneficial. They are proven to aid in training AI models to achieve a higher accuracy rate, such as in the *Improving Deep Learning using Generic Data Augmentation* scholarly paper by Luke Taylor and Geoff Nitschke, published in 2017 by the *Arxiv* website.

The code for the contrast, saturation, and hue-shifting wrapper functions is straightforward with the Albumentations library. Let's take a look:

```
# for contrast
aug_album = albumentations.ColorJitter(brightness=0.0,
    contrast=contrast, saturation=0.0,
    hue=0.0, always_apply=True, p=1.0)
# for saturation
aug_album = albumentations.ColorJitter(brightness=0.0,
    contrast=0.0, saturation=saturation,
    hue=0.0, always_apply=True, p=1.0)
# for hue shifting
aug_album = albumentations.ColorJitter(brightness=0.0,
    contrast=0.0, saturation=0.0,
    hue=hue, always_apply=True, p=1.0)
```

Pluto has exaggerated the *unsafe* value so that you can see the results in this book. The *unsafe* parameter for contrast in the sea animals dataset is as follows:

```
# use the contrast wrapper function on sea animal image
pluto.draw_image_contrast(pluto.df_sea_animal,
   contrast=8.5,
   bsize=2)
```

The output is as follows:

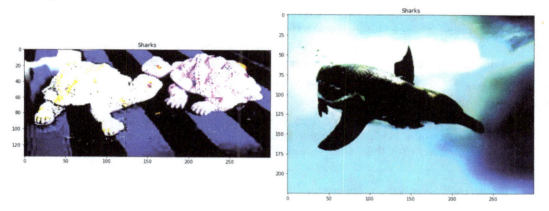

Figure 3.26 – Unsafe use of contrast on the sea animals dataset

The *unsafe* parameter for saturation in the food dataset is as follows:

```
# use the contrast wrapper function on food image
pluto.draw_image_saturation(pluto.df_food,
   saturation=10.5)
```

The output is as follows:

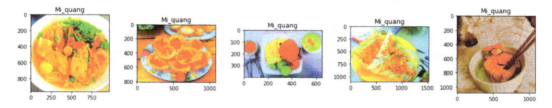

Figure 3.27 – Unsafe use of saturation on the food dataset

The *unsafe* parameter for hue shifting in the People dataset is as follows:

```
# use the contrast wrapper function on people photo
pluto.draw_image_hue(pluto.df_people,hue=0.5)
```

The output is as follows:

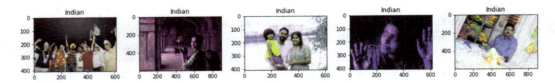

Figure 3.28 – Unsafe use of hue shifting on the People dataset

Contrast, saturation, and hue shifting apply to five of the image datasets, and the key is to find the safe range for each dataset. The exception is the medical images – the Covid-19 photos. You need to consult a domain expert to see how much you can distort the images and retain their integrity.

> **Fun challenge**
>
> Here is a thought experiment: can you think of an image category that would safely use hue shifting? In other words, in what photo subject can you shift the hue value and not compromise the image's integrity? Hint: think of an animal that hunts by sonar or heat source.

The next filter we'll cover is the noise injection filter, which can be easily recognized in this book's grayscale photos.

Noise injection

Noise injection is a strange filter because it is counterintuitive. Image augmentation distorts the original pictures within a safe limit, but injecting noise into a photo causes the images to degrade deliberately.

The scholarly paper *Data Augmentation in Training CNNs: Injecting Noise to Images*, by Murtaza Eren Akbiyik, published in 2019, and reviewed at the *ICLR 2020 Conference*, analyzes the effects of adding or applying different noise models of varying magnitudes to CNN architectures. It shows that noise injection provides a better understanding of optimal learning procedures for image classification.

For the `draw_image_noise()` wrapper method, Pluto uses the Albumentation's Gaussian noise method, as follows:

```
# use Albumentations for noise injection
aug_album = albumentations.GaussNoise(var_limit=var_limit,
    always_apply=True,
    p=1.0)
```

Pluto bumps the noise level to the extreme for the exaggerated *unsafe* case. The command in the Python Notebook is as follows:

```
# use noise wrapper function on fungi photo
pluto.draw_image_noise(pluto.df_fungi,
    var_limit=(10000.0, 20000.0), bsize=2)
```

The output is as follows:

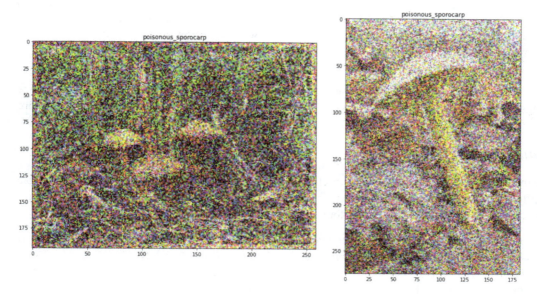

Figure 3.29 – Unsafe use of noise injection on the fungi dataset

Since the goal of the mall crowd dataset is to estimate the crowd size, adding some noise to the image is acceptable. Pluto found that the *safe* noise level is from about 200 to 400. The command is as follows:

```
# use noise wrapper function on crowd image
pluto.draw_image_noise(pluto.df_crowd,
    var_limit=(200.0, 400.0),
    bsize=2)
```

The result is as follows:

Figure 3.30 – Safe use of noise injection on the mall crowd dataset

> **Fun challenge**
>
> Here is both a thought and a hands-on experiment. Can you define a set of ranges for each image augmentation that applies to a specific image topic, such as landscape, birds, or house appliances? If you think that is possible, can you write a Python function that uses Pluto's wrapper functions?

This is when Pluto begins experimenting with more exotic filters, but he limits his choices to the image augmentation methods studied in published scholarly papers. The next two filters we will look at are the rain and sun flare filters.

Rain and sun flare

In image augmentation, rain and sun flare effects are not widely used in AI. However, it is an acceptable option if the image domain is landscape or cityscape. The rain and sun flare implementations are simplistic because they are optimized for speed over a realistic depiction of rain or sun flare.

If you require a natural rain effect, then you can refer to a paper that presents a new approach to synthesizing realistic rainy scenes using a **generative adversarial network (GAN)**: *Synthesized Rain Images for Deraining Algorithms*, by Jaewoong Choi, Dae Ha Kim, Sanghyuk Lee, Sang Hyuk Lee, and Byung Cheol Song, published in 2022 in *Neurocomputing Volume 492*.

The realistic rendering will take some time. Therefore, you should use the pre-processing augmentation method, which generates the images and saves them to local or cloud disk storage, before training the AI model.

Pluto does not have access to the GAN method, so he uses the Albumentations library for dynamically generating the effects. The key code inside the `draw_image_rain()`, and `draw_image_sunflare()` wrapper functions is as follows:

```
# for rain
aug_album = albumentations.RandomRain(
  drop_length = drop_length,
  drop_width=drop_width,
  blur_value=blur_value,
  always_apply=True,
  p=1.0)
# for sun flare
aug_album = albumentations.RandomSunFlare(
  flare_roi = flare_roi,
  src_radius=src_radius,
  always_apply=True, p=1.0)
```

Pluto exaggerates the effects of the sun flare filter to an *unsafe* level. The command is as follows:

```
# use sunflare wrapper function with people photo
pluto.draw_image_sunflare(pluto.df_people,
```

```
flare_roi=(0, 0, 1, 0.5),
src_radius=400,
bsize=2)
```

The output is as follows:

Figure 3.31 – Unsafe use of the sun flare filter on the People dataset

Pluto discovered that the *safe* level for the fungi dataset is a radius of 120, with a flare-roi of (0, 0, 1). The command is as follows:

```
# use the sunflare wrapper function on fungi image
pluto.draw_image_sunflare(pluto.df_fungi, src_radius=120)
```

The output is as follows:

Figure 3.32 – Safe use of the sun flare filter on the fungi dataset

For the People dataset, Pluto found that the safe parameter is a `drop_length` equal to 20, a `drop_width` equal to 1, and a `blur_value` equal to 1:

```
# use the rain wrapper function on people photo
pluto.draw_image_rain(pluto.df_people, drop_length=20,
    drop_width=1, blur_value=1)
```

The result is as follows:

Figure 3.33 – Safe use of the rain filter on the People dataset

Many more photometric transformations are available; for example, Albumentations has over 70 image filters. However, for now, Pluto will present two more effects: the Sepia and FancyPCA filters.

Sepia and FancyPCA

Sepia involves altering the color tone to a brownish color. This brown is the color of cuttlefish ink, and the result gives the effect of old or aged pictures. Fancy **Principal Components Analysis (FancyPCA)** color augmentation alters the RGB channels' intensities along the images' natural variations.

A scholarly research paper used the FancyPCA filter to improve DL prediction of rock properties in reservoir formations: *Predicting mineralogy using a deep neural network and fancy PCA* by Dokyeong Kim, Junhwan Choi, Dowan Kim, and Joongmoo Byun, in 2022, presented at the *SEG International Exposition and Annual Meeting*.

For the `draw_image_sepia()` and `draw_image_fancyPCA()` wrapper functions, Pluto uses the Albumentations library:

```
# for sepia use albumentations library
aug_album = albumentations.ToSepia(always_apply=True, p=1.0)
# for FancyPCA use albumentations library
aug_album = albumentations.FancyPCA(alpha=alpha,
    always_apply=True,
    p=1.0)
```

You can see the results in the Python Notebook's color output images. Pluto has chosen the People dataset to experiment with the sepia and FancyPCA filters because it has no objective. Assuming the target is to classify people's age ranges, both filters are applicable. For the sepia filter, the command is as follows:

```
# use the sepia wrapper function on people photo
pluto.draw_image_sepia(pluto.df_people)
```

The output is as follows:

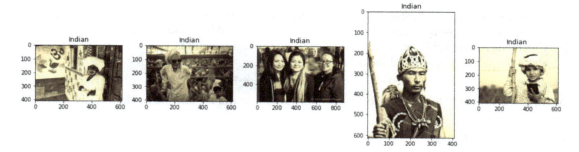

Figure 3.34 – Safe use of sepia on the People dataset

Pluto overstates the FancyPCA filter to an *unsafe* level by setting the alpha value to 5.0. The command is as follows:

```
# use fancyPCA wrapper function on people photo
pluto.draw_image_fancyPCA(pluto.df_people,alpha=5.0,bsize=2)
```

The result is as follows:

Figure 3.35 – Unsafe use of FancyPCA on the People dataset

So far, you and Pluto have covered and written many wrapper functions for photometric transformations, such as lighting, grayscale, contrast, saturation, hue shifting, noise injection, rain, sun flare, sepia, and FancyPCA. Still, there are far more image filters in the Albumentations library. Pluto follows the golden image augmentation rule for selecting a filter that improves prediction accuracy, as a published scholarly paper describes, such as *The Effectiveness of Data Augmentation in Image Classification using Deep Learning* paper by **Luis Perez, Jason Wang**, published by the *Cornell University Arxiv* in December 2017.

Fun challenge

Here is a thought experiment: there are too many image augmentation techniques to count. So, how do you know which augmentation function is suitable to use? For example, is the Cinematic Anamorphic photo filter an effective image augmentation technique? Hint: think about the subject domain and the processing speed.

Moving away from photographic transformations, next, Pluto will dig into the random erasing filter.

Random erasing

Random erasing adds a block of noise, while noise injection adds one pixel at a time.

Two recently published papers show that random erasing filters and extended random erasing increase the prediction accuracy of the DL model. The first paper is called *Random Erasing Data Augmentation*, by Zhun Zhong, Liang Zheng, Guoliang Kang, Shaozi Li, and Yi Yang, in 2020, and was presented at the *AAAI Conference on Artificial Intelligence*. The second paper is called *Perlin Random Erasing for Data Augmentation*, by Murat Saran, Fatih Nar, and Ayşe Nurdan Saran, in 2021, and was presented at the 29th *Signal Processing and Communications Applications Conference (SIU)*.

Pluto uses the Fast.ai library in the `draw_image_erasing()` wrapper function. The pertinent code is as follows:

```
# use fastai library for random erasing
aug = [fastai.vision.augment.RandomErasing(p=1.0,
  max_count=max_count)]
itfms = fastai.vision.augment.Resize(480)
```

It is challenging to find a safe level for random erasing. It depends on the image subject, DL base model, and target label. Generally, Pluto selects a random erasing safe parameter and trains the AI model. If the DL model is overfitting, then he increases the random erasing effects, and if the model's prediction accuracy is diverging, he decreases the random erasing parameters. Here is a safe starting point for the food dataset:

```
# use random erasing wrapper function on food image
pluto.dls_food = pluto.draw_image_erasing(
  pluto.df_food,
```

```
bsize=6,
max_count=4)
```

The output is as follows:

Figure 3.36 – Unsafe use of FancyPCA on the food dataset

So far, Pluto uses one filter at a time. Next, he will combine multiple geographic and photographic transformations with random erasing.

Combining

The power of image augmentation is that Pluto can combine multiple image filters in one dataset. This increases the number of images for training by a multiplication factor.

Fun fact

The horizontal flip filter's default is set to 50% probability. The result is that the image size increases by half – that is, `total_flip = image_size + (0.5 * image_size)`. The image size will increase by a multiplication factor when the random cropping and padding are added together with a padding mode of 3 – that is, `total_2_combine = total_fip + (3 * (image_size + (0.5 * image_size)) + (image_size * random_croping_factor))`, where `random_croping_factor` is between zero and the safe cropping value, which is less than 1.0.

In this chapter, Pluto covered 15 image augmentation methods. Therefore, combining most or all of the filters into one dataset will increase its size substantially. Increasing the total number of images for training in DL is a proven method to reduce or eliminate the overfitting problem.

There are general rules for the applicable filters and safe parameters that should work with most image datasets. However, Pluto follows the golden rule of image augmentation. This golden rule selects which image filter to use and sets the safe parameters based on the photo subject and the predictive model's goal. In other words, each project will have different image augmentation filters and safe parameters.

Before unveiling the table representing the safe parameters for each filter per six real-world image datasets, Pluto must review the image datasets' topics and goals:

- The **Covid-19** dataset consists of people's chest X-ray images. The goal is to predict between Covid-19, viral pneumonia, and normal.

- The **People** dataset consists of pictures of everyday people. No goal is stated, but Pluto assumes the usage could classify age, sex, ethnicity, emotional sentiment, and facial recognition.

- The **Fungi** dataset consists of photos of fungi in a natural environment. The goal is to predict if the fungi are edible or poisonous.

- The **Sea Animal** dataset consists of images of sea animals, mainly underwater. The goal is to classify the 19 sea animals provided.

- The **Food** dataset consists of images of commonly served Vietnamese dishes. The goal is to classify the 31 types of dishes.

- The **Mall Crowd** dataset consists of images of people in a typical shopping mall. The goal is to predict the size of the crowd.

To generate the filters and safe parameters table for each image dataset, Pluto has written a quick function, `print_safe_parameters()`, using pandas, because he thinks coding is fun. For readability, there are two parts to the table, as follows:

	Filter	Covid-19	People	Fungi
0	Horizontal Flip	NA	Yes	Yes
1	Vertical Flip	NA	NA	NA
2	Croping and Padding	NA	pad=border	pad=border
3	Rotation	NA	max_rotate=25.0	max_rotate=25.0
4	Warping	NA	magnitude=0.3	magnitude=0.3
5	Lighting	brightness=0.2	brightness=0.3	brightness=0.3
6	Grayscale	NA	NA	NA
7	Contrast	contrast=0.1	contrast=0.3	contrast=0.3
8	Saturation	NA	saturation=3.5	saturation=2.0
9	Hue Shifting	NA	NA	NA
10	Noise Injection	limit=(100.0, 300.0)	limit=(300.0, 500.0)	limit=(200.0, 400.0)
11	Sun Flare	NA	NA	radius=200
12	Rain	NA	NA	length=20
13	Sepia	NA	Yes	NA
14	FancyPCA	NA	alpha=0.5	alpha=0.5
15	Random Erasing	NA	max_count=3	max_count=3

Figure 3.37 – Safe parameter for each image dataset – part 1

Figure 3.37 shows the first half of the big table, and *Figure 3.38* displays the second half.

	Filter	Sea Animal	Food	Mall Crowd
0	Horizontal Flip	Yes	Yes	Yes
1	Vertical Flip	Yes	Yes	NA
2	Croping and Padding	pad=reflection	pad=reflection	pad=zeros
3	Rotation	max_rotate=180.0	max_rotate=180.0	max_rotate=16.0
4	Warping	magnitude=0.4	magnitude=0.4	magnitude=0.3
5	Lighting	brightness=0.4	brightness=0.4	brightness=0.3
6	Grayscale	NA	NA	Yes
7	Contrast	contrast=0.3	contrast=0.4	contrast=0.4
8	Saturation	saturation=3.0	saturation=3.0	saturation=2.5
9	Hue Shifting	hue=0.15	hue=0.2	hue=0.2
10	Noise Injection	limit=(200.0, 400.0)	limit=(300.0, 400.0)	limit=(300.0, 500.0)
11	Sun Flare	NA	NA	NA
12	Rain	NA	NA	NA
13	Sepia	NA	NA	NA
14	FancyPCA	alpha=0.5	alpha=0.5	NA
15	Random Erasing	max_count=4	max_count=4	NA

Figure 3.38 – Safe parameter for each image dataset – part 2

The safe parameters are from Pluto's exploration of the Python Notebook, but you may find more suitable values than Pluto. There are no rigid or fixed rules regarding image augmentation. Therefore, you should use the Python Notebook to explore the possibilities. If you read a scholarly paper about a new image augmentation technique, implement it using the Albumentations or other image libraries.

Pluto has written six wrapper functions for reinforcing learning through coding, one for each image dataset.

> **Fun fact**
> You can run the wrapper function repeatedly because it generates a different image set every time. In addition, it will randomly select other base images from the real-world dataset. Therefore, you can run it a thousand times and only see the same output once.

Each wrapper function defines a set of Fast.ai image augmentation filters; for example:

```
# use fastai library for brightness and contrast
aug = [
  fastai.vision.augment.Brightness(max_lighting=0.3,p=0.5),
  fastai.vision.augment.Contrast(max_lighting=0.4, p=0.5)]
# use albumentation library
albumentations.Compose([
  albumentations.GaussNoise(var_limit=(100.0, 300.0),
    p=0.5)])
```

In addition, the wrapper function uses a helper method to fetch the `Albumentations` filters – for example, `_fetch_album_covid19()`.

Pluto reviews the image augmentation for the Covid-19 dataset by using the following command in the Python Notebook:

```
# use covid 19 wrapper function
pluto.draw_augment_covid19(pluto.df_covid19)
```

The output is as follows:

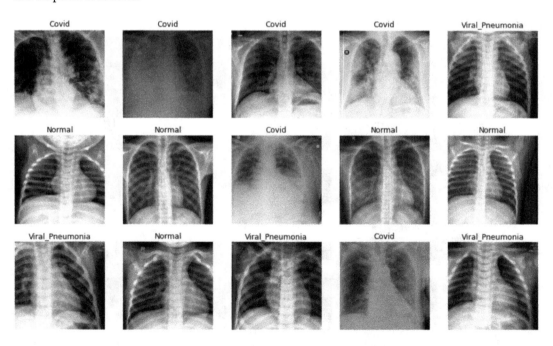

Figure 3.39 – Image augmentation for the Covid-19 dataset

The relevant code lines for the combination filters for the People dataset are as follows:

```
# use both fastai and albumentation library
aug = [
  fastai.vision.augment.Brightness(max_lighting=0.3,p=0.5),
  fastai.vision.augment.Contrast(max_lighting=0.4, p=0.5),
  AlbumentationsTransform(self._fetch_album_covid19())]
# use alpbumentation library in the _fetch_albm_covid()
albumentations.Compose([
  albumentations.ColorJitter(brightness=0.3,
    contrast=0.4, saturation=3.5,hue=0.0, p=0.5),
  albumentations.ToSepia(p=0.5),
  albumentations.FancyPCA(alpha=0.5, p=0.5),
  albumentations.GaussNoise(var_limit=(300.0, 500.0), p=0.5)
  ])
```

Pluto reviews the image augmentation for the People dataset by using the following command in the Python Notebook:

```
# use people wrapper function
pluto.draw_augment_people(pluto.df_people)
```

The output is as follows:

Figure 3.40 – Image augmentation for the People dataset

The relevant code lines for the fungi combination filters are as follows:

```python
# use both fastai and albumentations libraries
aug = [
  fastai.vision.augment.Flip(p=0.5),
  fastai.vision.augment.Rotate(25.0,p=0.5,pad_mode='border'),
  fastai.vision.augment.Warp(magnitude=0.3,
    pad_mode='border',p=0.5),
  fastai.vision.augment.RandomErasing(p=0.5,max_count=2),
  AlbumentationsTransform(self._fetch_album_fungi())]
# use albumentation inside the _fetch_album_fungi()
albumentations.Compose([
  albumentations.ColorJitter(brightness=0.3,
    contrast=0.4, saturation=2.0,hue=0.0, p=0.5),
  albumentations.FancyPCA(alpha=0.5, p=0.5),
  albumentations.RandomSunFlare(flare_roi=(0, 0, 1, 0.5),
    src_radius=200, always_apply=True, p=0.5),
  albumentations.RandomRain(drop_length=20,
    drop_width=1.1,blur_value=1.1,always_apply=True, p=0.5),
  albumentations.GaussNoise(var_limit=(200.0, 400.0), p=0.5)
  ])
```

Pluto reviews the image augmentation for the fungi dataset by using the following command:

```python
# use fungi wrapper function
pluto.draw_augment_fungi(pluto.df_fungi)
```

The output is as follows:

Figure 3.41 – Image augmentation for the fungi dataset

The relevant code lines for the sea animal combination filters are as follows:

```
# use both fastai and albumentations library
aug = [
  fastai.vision.augment.Dihedral(p=0.5,
    pad_mode='reflection'),
  fastai.vision.augment.Rotate(180.0,
    p=0.5,pad_mode='reflection'),
  fastai.vision.augment.Warp(magnitude=0.3,
    pad_mode='reflection',p=0.5),
  fastai.vision.augment.RandomErasing(p=0.5,max_count=2),
  AlbumentationsTransform(self._fetch_album_sea_animal())]
# use albumentations for _fetch_album_sea_animal()
albumentations.Compose([
  albumentations.ColorJitter(brightness=0.4,
    contrast=0.4, saturation=2.0,hue=1.5, p=0.5),
  albumentations.FancyPCA(alpha=0.5, p=0.5),
  albumentations.GaussNoise(var_limit=(200.0, 400.0),
    p=0.5)])
```

Pluto reviews the image augmentation for the sea animal dataset by using the following command in the Python Notebook:

```
# use the sea animal wrapper function
pluto.draw_augment_sea_animal(pluto.df_sea_animal)
```

The output is as follows:

Figure 3.42 – Image augmentation for the sea animals dataset

The relevant code lines for the food combination filters are as follows:

```
# use both fastai and albumentations libraries
aug = [
  fastai.vision.augment.Dihedral(p=0.5,
    pad_mode='reflection'),
  fastai.vision.augment.Rotate(180.0,
    p=0.5,pad_mode='reflection'),
  fastai.vision.augment.Warp(magnitude=0.3,
    pad_mode='reflection',p=0.5),
  fastai.vision.augment.RandomErasing(p=0.5,max_count=2),
  AlbumentationsTransform(self._fetch_album_food())]
# use albumentation library for _fetch_album_food()
albumentations.Compose([
  albumentations.ColorJitter(brightness=0.4,
```

```
        contrast=0.4, saturation=2.0,hue=1.5, p=0.5),
    albumentations.FancyPCA(alpha=0.5, p=0.5),
    albumentations.GaussNoise(var_limit=(200.0, 400.0),
      p=0.5)])
```

Pluto reviews the image augmentation for the food dataset by using the following command:

```
# use food wrapper function
pluto.draw_augment_food(pluto.df_food)
```

The output is as follows:

Figure 3.43 – Image augmentation for the food dataset

The relevant code lines for the mall crowd combination filters are as follows:

```
# use both fastai and albumentations libraries
aug = [
  fastai.vision.augment.Flip(p=0.5),
  fastai.vision.augment.Rotate(25.0,
    p=0.5,pad_mode='zeros'),
  fastai.vision.augment.Warp(magnitude=0.3,
    pad_mode='zeros',p=0.5),
  fastai.vision.augment.RandomErasing(p=0.5,max_count=2),
  AlbumentationsTransform(self._fetch_album_crowd())]
```

```
# use albumentation library for _fetch_album_crowd()
albumentations.Compose([
  albumentations.ColorJitter(brightness=0.3,
    contrast=0.4, saturation=3.5,hue=0.0, p=0.5),
  albumentations.ToSepia(p=0.5),
  albumentations.FancyPCA(alpha=0.5, p=0.5),
  albumentations.GaussNoise(var_limit=(300.0, 500.0),
    p=0.5)])
```

Pluto reviews the image augmentation for the mall crowd dataset by using the following command:

```
# use the crowd wrapper function
pluto.draw_augment_crowd(pluto.df_crowd)
```

The output is as follows:

Figure 3.44 – Image augmentation for the mall crowd dataset

Every time Pluto runs any of the draw augmentation methods, there is an equal chance that one of the filters will be selected and that 50% of the filter will be applied per image in the batch. Pluto can override the default batch size of 15 using the `bsize` parameter. Since Pluto employs the safe range on all filters, you may not notice the difference. However, that is expected because the goal is to distort the images without compromising the target label before pre-processing.

> **Fun challenge**
> Write a new combination wrapper function for a real-world dataset. If you have not downloaded or imported a new real-world image dataset, do so now and write a combined wrapper function as Pluto did.

This was a challenging chapter. Together, you and Pluto learned about image augmentation and how to use Python code to gain a deeper insight. Now, let's summarize this chapter.

Summary

In the first part of this chapter, you and Pluto learn about the image augmentation concepts for classification. Pluto grouped the filters into geometric transformations, photometric transformations, and random erasing to make the image filters more manageable.

When it came to geometric transformations, Pluto covered horizontal and vertical flipping, cropping and padding, rotating, warping, and translation. These filters are suitable for most image datasets, and there are other geometric transformations, such as tilting or skewing. Still, Pluto followed the golden image augmentation rule for selecting a filter that improves prediction accuracy described in a published scholarly paper.

This golden rule is more suitable for photometric transformations because there are about 70 image filters in the Albumentations library and hundreds more available in other image augmentation libraries. This chapter covered the most commonly used photometric transformations cited in published scholarly papers: lighting, grayscale, contrast, saturation, hue shifting, noise injection, rain, sepia, and FancyPCA. You are encouraged to explore more filters from the Albumations library in the Python Notebook provided.

The second part of this chapter consisted of Python code to reinforce your understanding of various image augmentation techniques. Pluto led you through the process of downloading the six real-world image datasets from the *Kaggle* website. Pluto wrote the fetching data code in *Chapter 2*. He reused the fetch functions to retrieve the Covid-19, people, fungi, sea animals, food, and mall crowd real-world image datasets.

Before digging into the code, Pluto reviewed seven popular image augmentation libraries: Albumentations, Fast.ai, PIL, OpenCV, scikit-learn, Mahotas, and GraphicsMagick. Pluto used the Albumentations, Fast.ai, and PIL libraries to code the wrapper functions.

The goal of these wrapper functions was to explain each image filter clearly. In all cases, the functions use the library augmentation methods under the hood. Many photometric transformations are more visible in the Python Notebook's color output. Pluto showed the safe and *unsafe* parameters for each filter applied to the six image datasets. You are encouraged to run and even hack the Python Notebook's code because there are no absolute right or wrong safe levels.

A lot of Python code was provided, but it consisted of simple wrapper functions that followed good OOP standards and there were no other methods. The goal was to give you insight into each image filter and make it easy for you to explore and hack the code provided.

At the end of this chapter, Pluto pulled together to create an image filter combination table customized for each of the six image datasets. He then wrote six combined augmentation methods for each image dataset.

Throughout this chapter, there were *fun facts* and *fun challenges*. Pluto hopes you will take advantage of these and expand your experience beyond the scope of this chapter.

In the next chapter, we will cover image segmentation, which reuses many of the image classification functions that were covered in this chapter.

4

Image Augmentation
for Segmentation

Image segmentation, like image classification, is the cornerstone in the computer vision domain. Image segmentation involves grouping parts of an image that belong to the same object, also known as pixel-level classification. Unlike image classification, which identifies and predicts the subject or label of a photo, image segmentation determines whether a pixel belongs to a list of objects – for example, an urban photograph has streets, street signs, cars, trucks, bicycles, buildings, trees, and pedestrians. Image segmentation's job is to decide whether this image pixel belongs to a car, tree, or other objects.

Deep learning (**DL**), an **artificial neural network** (**ANN**) algorithm, has made a tremendous breakthrough in image segmentation. For example, image segmentation in DL makes it possible for autonomous vehicles and **Advanced Driver Assistance Systems** (**ADASes**) to detect navigable surfaces or pedestrians. Many medical applications use segmentation for tumor boundary drawing or measuring tissue volumes, for example.

The image augmentation methods for segmentation or classification are the same, except segmentation comes with an additional mask image or ground-truth image. Therefore, most of what we learned about augmenting images for classification in *Chapter 3* applies to augmenting segmentation.

This chapter aims to provide continuing geometric and photometric transformations for image segmentation. In particular, you will learn about the following topics:

- Geometric and photometric transformations
- Real-world segmentation datasets
- Reinforcing your learning through Python code

> **Fun fact**
>
> Image segmentation or semantic segmentation is used in many self-driving car AI controllers. It is used to identify objects and people on a street. Worldwide competition wins or losses primarily due to image segmentation augmentation techniques, such as the *Udacity and Lyft Perception Challenge* winner of the *Kaggle* competition, use random resized crop, horizontal flip, and random color jitter in brightness, contrast, and saturation.

Let's begin with the geometric and photometric transformations for segmentation.

Geometric and photometric transformations

As discussed in *Chapter 3*, geometric transformations alter a picture's geometry, such as by flipping, cropping, padding, rotating, or resizing it. For segmentation, when horizontally **flipping** an image, the same must be done for the mask. Pluto will show you how to flip an original and accompanying mask image; here is a sneak peek:

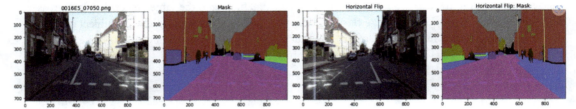

Figure 4.1 – Image segmentation horizontal flip

Many of the **safe** values discussed in *Chapter 3* stay mostly the same. For example, if the picture's subject is people or an urban cityscape, the classification augmentation can't flip vertically because the prediction of people's age or the city's name relies on the picture not being upside down. However, segmentation aims to group or draw an outline of the people or cars. Thus, vertical flipping is acceptable.

The safe range needs further investigation for many real-world applications. For example, for self-driving automobiles, what if you are in a car accident and your vehicle is upside down? Does the AI still need to classify its surroundings correctly?

Photometric transformations, such as brightness, saturation, contrast, hue shifting, and FancyPCA, are more problematic to apply to segmentation because the original image is distorted but not the mask image. The big question is, would augmenting the original but not the mask image increase the prediction's accuracy?

Noise injection, **random erasing**, **snow**, and **rain** transformations are not applicable to segmentation because they introduce new pixels. The mask image can't compensate for the replacement pixels. Similarly, blurring or embossing filters are not suitable for segmentation. In the `Albumentations` library, 37 transformations are defined as safe for distorting both original and mask images.

Technically, you can use photometric transformations for segmentation with Python code, but it is wise to research published scholarly papers for confirmation. The golden augmentation rule that we discussed in *Chapter 3* is applied here as well – you select a filter that improves the prediction accuracy described in a published academic paper.

Learning by using Python code is another angle you can use to understand image segmentation. However, before we do that, let's ask Pluto to download a few real-world segmentation datasets from Kaggle.

Real-world segmentation datasets

The *Kaggle* website is an online community platform for data scientists and ML devotees. It contains thousands of real-world datasets, as mentioned in *Chapters 1, 2*, and *3*.

When searching for image segmentation datasets, Pluto found about 500 useable real-world segmentation datasets. The topics range from self-driving automobiles and medicine to micro-fossils. Pluto picked two segmentation datasets from popular market segments.

The other consideration is that the image type must be easy to work with in the Albumentations library. Pluto uses the **PIL** and **NumPy** libraries to read and convert the photos into a three-dimensional array. The original image's **shape** is (width, height, and depth), where depth is usually equal to three. The mask image's **shape** is (width, height), where the value is 0, 1, 2, and so on up to the number of labels.

> **Fun fact**
>
> The PIL library can read image formats such as `.jpg`, `.gif`, `.tiff`, `.png`, and about 50 other image formats. Still, sometimes, the real-world segmentation datasets come with an image format that PIL can't read. In those cases, Pluto relies on the Python **ImageIO** library, which can read over 100 image types.

The two selected segmentation datasets are as follows:

- The *Cambridge-Driving Labeled Video (CamVid)* database is the first real-world segmentation dataset. The context on the *Kaggle* website is as follows:

  "*The Cambridge-Driving Labeled Video Database (CamVid) provides ground truth labels that associate each pixel with one of 32 semantic classes. This dataset is often used in (real-time) semantic segmentation research.*"

  It was published in 2020 by the **University of Cambridge**, and the license is **CC BY-NC-SA 4.0**: `https://creativecommons.org/licenses/by-nc-sa/4.0/`.

- The second real-world dataset is called *Semantic segmentation of aerial imagery*. The description from the *Kaggle* website is as follows:

  "*The dataset consists of aerial imagery of Dubai obtained by MBRSC satellites and annotated with pixel-wise semantic segmentation in 6 classes. The total volume of the dataset is 72 images grouped into 6 larger tiles.*"

  It was published in 2020 by the **Roia Foundation in Syria**, and the license is **CC0: Public Domain**: https://creativecommons.org/publicdomain/zero/1.0/.

After selecting the two segmentation datasets, the following four steps should be familiar to you by now. Review *Chapters 2* and *3* if you need clarification. The steps are as follows:

1. Retrieve the Python Notebook and Pluto.

2. Download real-world data.

3. Load the data into pandas.

4. View the data images.

Fun challenge

Find and download two additional image segmentation datasets from the *Kaggle* website or other sources. *Kaggle* competitions and data consist of hundreds of image segmentation datasets. Thus, finding image segmentation datasets that are meaningful to you or your job shouldn't be challenging. Hint: use Pluto's `fetch_kaggle_dataset()` or `fetch_kaggle_comp_data()` function.

Let's start with Pluto.

Python Notebook and Pluto

Start by loading the `data_augmentation_with_python_chapter_4.ipynb` file into Google Colab or your chosen Jupyter Notebook or JupyterLab environment. From this point onward, the code snippets will be from the Python Notebook, which contains the complete functions.

Next, you must clone the repository and use the `%run` command to start Pluto:

```
# clone the github
!git clone 'https://github.com/PacktPublishing/Data-
Augmentation-with-Python'
# instantiate Pluto
%run 'Data-Augmentation-with-Python/pluto/pluto_chapter_3.py'
```

The output will be as follows or similar:

```
----------------------------- : -----------------------------
        Hello from class : <class '__main__.PacktDataAug'> Class:
PacktDataAug
              Code name : Pluto
              Author is : Duc Haba
----------------------------- : -----------------------------
            fastai 2.6.3 : actual 2.7.9
----------------------------- : -----------------------------
      albumentations 1.2.1 : actual 1.2.1
----------------------------- : -----------------------------
```

Double-check that Pluto has loaded correctly:

```
# display Python and libraries version number
pluto.say_sys_info()
```

The output will be as follows or something similar, depending on your system:

```
----------------------------- : -----------------------------
            System time : 2022/10/21 15:46
               Platform : linux
    Pluto Version (Chapter) : 3.0
      Python version (3.7+) : 3.7.13 (default, Apr 24 2022, 01:04:09)
[GCC 7.5.0]
        PyTorch (1.11.0) : actual: 1.12.1+cu113
          Pandas (1.3.5) : actual: 1.3.5
             PIL (9.0.0) : actual: 7.1.2
      Matplotlib (3.2.2) : actual: 3.2.2
             CPU count : 2
             CPU speed : NOT available
----------------------------- : -----------------------------
```

Pluto has verified that the Python Notebook is working correctly. The next step is downloading real-world image datasets from Kaggle.

Real-world data

The following download function is from *Chapter 2*. Pluto has reused this here:

```
# Fetch Camvid photo
url = 'https://www.kaggle.com/datasets/carlolepelaars/camvid'
pluto.fetch_kaggle_dataset(url)
# Fetch Aerial image
```

```
url = 'https://www.kaggle.com/datasets/humansintheloop/semantic-
segmentation-of-aerial-imagery'
pluto.fetch_kaggle_dataset(url)
```

Before viewing the downloaded photos, Pluto needs to load the information into a pandas DataFrame.

Pandas

A few cleanup tasks need to be done here, such as replacing a space character with an underscore character in the directories or filenames and separating original and mask images. After the cleanup, Pluto reuses the `make_dir_dataframe()` function to read the original image data into a pandas DataFrame. The command for the CamVid data is as follows:

```
# import data to Pandas
f = 'kaggle/camvid/CamVid/train'
pluto.df_camvid = pluto.make_dir_dataframe(f)
```

The output of the first three records is as follows:

	fname	label
0	kaggle/camvid/CamVid/train/0016E5_07590.png	train
1	kaggle/camvid/CamVid/train/Seq05VD_f03000.png	train
2	kaggle/camvid/CamVid/train/0016E5_08063.png	train

Figure 4.2 – CamVid pandas DataFrame, first three rows

The mask images are in a different folder, and the mask image's name has _L appended to the filename.

The primary reason for Pluto using pandas is that adding a new column for the matching mask and original filename is a trivial task. There are only two key code lines. The first is in the helper function to generate the correct mask image path, while the second is to create a new column for applying the helper function. The code for this is as follows:

```
# define helper function
@add_method(PacktDataAug)
def _make_df_mask_name(self,fname):
  p = pathlib.Path(fname)
  return (str(p.parent.parent) +
    '/' + str(p.parent.name) + '_labels/' +
    str(p.stem) + '_L' + str(p.suffix))
# method definition
@add_method(PacktDataAug)
def make_df_mask_name(self,df):
```

```
df['mask_name'] = df.fname.apply(self._make_df_mask_name)
return
```

The command to complete the CamVid DataFrame is as follows:

```
# create mask file name
pluto.make_df_mask_name(pluto.df_camvid)
```

The output is as follows:

	fname	mask_name
0	kaggle/camvid/CamVid/train/0016E5_02070.png	kaggle/camvid/CamVid/train_labels/0016E5_02070_L.png
1	kaggle/camvid/CamVid/train/0016E5_07999.png	kaggle/camvid/CamVid/train_labels/0016E5_07999_L.png
2	kaggle/camvid/CamVid/train/0016E5_08079.png	kaggle/camvid/CamVid/train_labels/0016E5_08079_L.png

Figure 4.3 – Complete CamVid DataFrame, first three rows

Once Pluto has gathered all the information squared away in the DataFrame, the next step is to display the original and mask images.

Viewing data images

Pluto could reuse the `draw_batch()` function from *Chapter 2* to display the original and mask images in separate batches, but the result does not reinforce the combination of original and mask images. Therefore, Pluto will hack the `draw_batch()` method and create a new `draw_batch_segmentation()` and a helper function called `_draw_batch_segmentation()`.

The result shows the original image, then the mask image, and repeats this process. The command for displaying the CamVid segmentation photos is as follows:

```
# use new batch display method for segmentation
pluto.draw_batch_segmentation(pluto.df_camvid,
  is_shuffle=True)
```

The output is as follows:

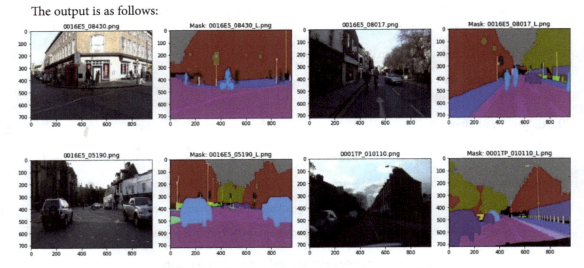

Figure 4.4 – CamVid's original and mask image batch

The segmentation batch looks correct, so Pluto repeats the same process for the aerial segmentation data.

Download the data with the following command:

```
# fetch real-world data
url = 'https://www.kaggle.com/datasets/humansintheloop/semantic-
segmentation-of-aerial-imagery'
pluto.fetch_kaggle_dataset(url)
```

Clean the directory and filenames, then import them into pandas with the following command:

```
# import to Pandas
f = 'kaggle/semantic-segmentation-of-aerial-imagery'
pluto.df_aerial = pluto.make_dir_dataframe(f)
```

The output for the first five records is as follows:

	fname	label
1	kaggle/semantic-segmentation-of-aerial-imagery/Semantic_segmentation_dataset/Tile_7/images/image_part_001.jpg	images
0	kaggle/semantic-segmentation-of-aerial-imagery/Semantic_segmentation_dataset/Tile_7/images/image_part_005.jpg	images
2	kaggle/semantic-segmentation-of-aerial-imagery/Semantic_segmentation_dataset/Tile_7/images/image_part_003.jpg	images

Figure 4.5 – Aerial pandas DataFrame, first three rows

Add the mask's filename using the new help function, as follows:

```
# create mask filename
pluto.make_df_mask_name_aerial(pluto.df_aerial)
```

The output for the first three records is as follows:

	mask_name
0	kaggle/semantic-segmentation-of-aerial-imagery/Semantic_segmentation_dataset/Tile_7/masks/image_part_005.png
2	kaggle/semantic-segmentation-of-aerial-imagery/Semantic_segmentation_dataset/Tile_7/masks/image_part_003.png
1	kaggle/semantic-segmentation-of-aerial-imagery/Semantic_segmentation_dataset/Tile_7/masks/image_part_001.png

Figure 4.6 – Complete aerial DataFrame, first three rows

Display the segmentation image batch with the following command:

```
# draw batch image
pluto.draw_batch_segmentation(pluto.df_aerial,
    is_shuffle=True)
```

The output is as follows:

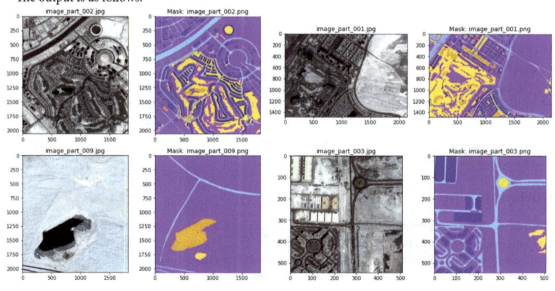

Figure 4.7 – Aerial pandas DataFrame, first five rows

> **Fun challenge**
>
> Here is a thought experiment: given an image dataset, how do you create the mask for the photos? Hint: you could use fancy image software to auto-trace the objects or outlines, then label them. The other options are Mechanical Turk or crowd-sourced. You should think about cost versus time.

Pluto has successfully downloaded and displayed the CamVid and aerial segmentation photos. Now, let's do some image augmentation with Python.

Reinforcing your learning

The same concepts for classification image transformations apply to segmentation image transformations. Here, Pluto reuses or slightly hacks the wrapper functions in *Chapter 3*. In particular, Pluto hacks the following methods for segmentation:

- Horizontal flip
- Vertical flip
- Rotating
- Random resizing and cropping
- Transpose
- Lighting
- FancyPCA

> **Fun fact**
>
> You can't complete or understand this chapter unless you have read *Chapter 3*. This is because Pluto reuses or slightly modifies the existing image augmentation wrapper functions.

Pluto chose these filters because the Albumentations library marked them as **safe** for segmentation. So, let's start with horizontal flip.

Horizontal flip

Pluto demonstrated horizontal flip using the PIL library in *Chapter 3* because the code is easy to understand. Thus, he will hack `draw_image_flip_pil()` into the `draw_image_flip_pil_segmen()` function. The transformation code is the same – that is, `PIL.ImageOps.mirror(img)`. The change is to display the images next to each other.

The command for flipping an image in the CamVid dataset in the Python Notebook is as follows:

```
# use wrapper function to flip image
pluto.draw_image_flip_pil_segmen(pluto.df_camvid.fname[0])
```

The output is as follows:

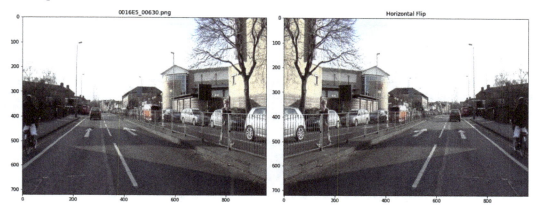

Figure 4.8 – Flipping an image using the PIL library

Pluto uses the same function for the mask image and passes the `mask_image` column into the pandas DataFrame. It is that easy. Pluto has to transform the original and mask images with the same filter.

The command for flipping the mask image is as follows:

```
# use wrapper function to flip image mask
pluto.draw_image_flip_pil_segmen(pluto.df_camvid.mask_name[0]
```

The output is as follows:

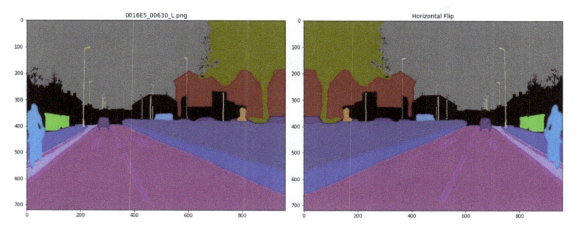

Figure 4.9 – Flipping the mask using the PIL library

> **Fun fact**
>
> Pluto only shows relevant code snippets in this book, but the fully functional object-oriented methods can be found in the Python Notebook. The code for this chapter looks remarkably similar to the code for *Chapter 3*. Pluto designed the software architecture using the principle layout provided in *Chapter 1*. Thus, the code looks clean but contains high complexity under the hood.

Under the hood, a color image is a three-dimensional array or a **Rank 3 tensor**. The image's shape is (width, height, and depth), where depth is usually equal to three, while the mask image's shape is (width, height), where the value is 0, 1, 2, and so on up to the number of labels. Therefore, mirroring a **Rank 3 tensor** follows the same operation as mirroring a **Rank 1 tensor**.

For the Albumentations library, the wrapper function for segmentation is as simple as the one provided in *Chapter 3*. The code for the `draw_image_flip_segmen()` method is as follows:

```python
# method definition
@add_method(PacktDataAug)
def draw_image_flip_segmen(self,df):
  aug_album = albumentations.HorizontalFlip(p=1.0)
  self._draw_image_album_segmentation(df,aug_album,
    'Horizontal Flip')
  return
```

It is the same as the `draw_image_flip()` function that we provided in *Chapter 3*. The difference is that a different helper function is used. Instead of using the `_draw_image_album()` helper function, it uses the `_draw_image_album_segmentation()` method.

The command for performing a horizontal flip on the CamVid segmentation data in the Python Notebook is as follows:

```python
# use wrapper function to flip both image and image mask
pluto.draw_image_flip_segmen(pluto.df_camvid)
```

The output is as follows:

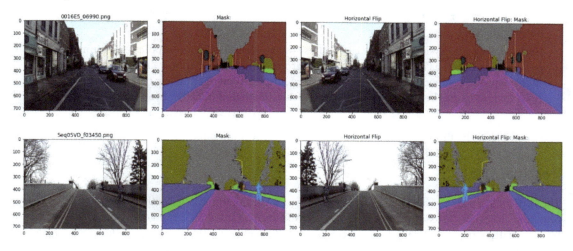

Figure 4.10 – Horizontal flip on the CamVid dataset

The command for performing a horizontal flip on the aerial segmentation data is as follows:

```
# use the same flip segmentation wrapper function on arial
pluto.draw_image_flip_segmen(pluto.df_aerial)
```

The output is as follows:

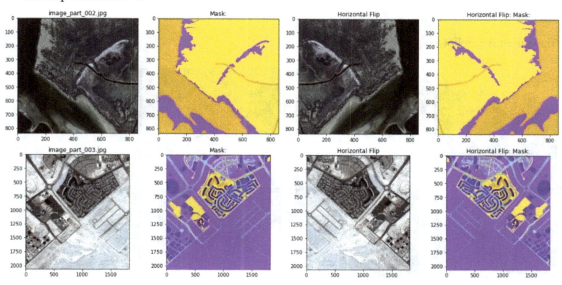

Figure 4.11 – Horizontal flip on the aerial dataset

Like in *Chapter 3*, the wrapper functions in this chapter randomly select a new image batch every time.

> **Fun challenge**
>
> Here is a thought experiment: how can you use image segmentation to support environmental organizations such as a wildlife conservation group? Hint: consider how customs agents can spot people selling parts of an endangered species, such as elephant ivory or saga horn, in an open market using their iPhones or **Close-Circuit Television** (**CCTV**) monitoring system.

Pluto completes the flipping transformation with the vertical flip filter.

Vertical flip

The vertical flip wrapper function is almost the same as the horizontal flip method. Pluto could write one uber function instead of each wrapper method individually. Still, the goal is to explain each transformation, not refactor it into more compact or efficient code. The key code line for the wrapper function is as follows:

```
# use albumentations library function
aug_album = albumentations.Flip(p=1.0)
```

The command for performing a vertical flip on the CamVid segmentation data in the Python Notebook is as follows:

```
# use flip wrapper function for camvid data
pluto.draw_image_flip_both_segmen(pluto.df_camvid)
```

The output is as follows:

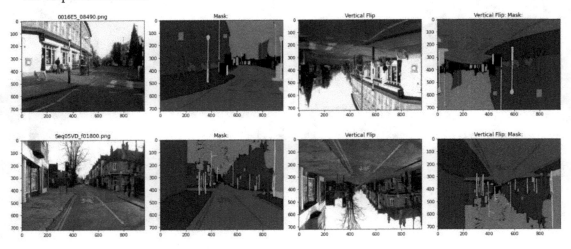

Figure 4.12 – Vertical flip on the CamVid dataset

The command for performing a vertical flip on the aerial segmentation data is as follows:

```
# use flip wrapper function for aerial image
pluto.draw_image_flip_both_segmen(pluto.df_aerial)
```

The output is as follows:

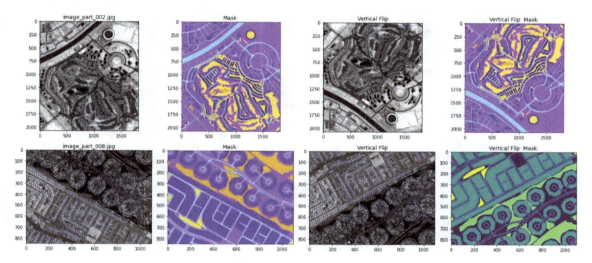

Figure 4.13 – Vertical flip on the aerial dataset

That concludes flipping. Now, let's look at rotating.

Rotating

The rotating safe parameter can go 45 degrees clockwise or counter-clockwise in direction. The Albumentations method is as follows:

```
# use albumentation library function for rotating
aug_album = albumentations.Rotate(limit=45, p=1.0)
```

The command for rotating the CamVid segmentation data in the Python Notebook is as follows:

```
# use rotate wrapper function for camvid image
pluto.draw_image_rotate_segmen(pluto.df_camvid)
```

The output is as follows:

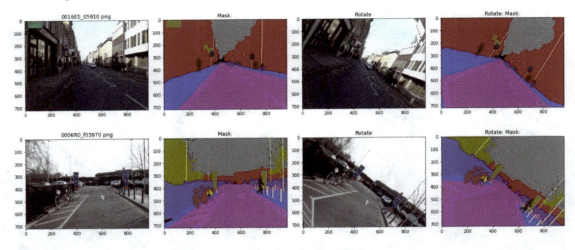

Figure 4.14 – Rotating the CamVid dataset

The command for rotating the aerial segmentation data is as follows:

```
# use rotate wrapper function for aerial image
pluto.draw_image_rotate_segmen(pluto.df_aerial)
```

The output is as follows:

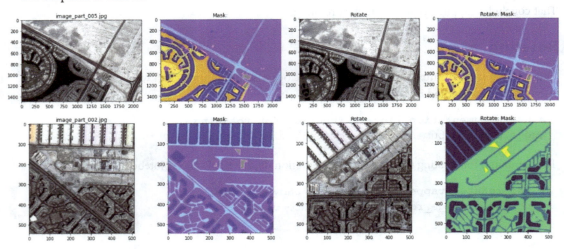

Figure 4.15 – Rotating the aerial dataset

The next filter is resizing and cropping.

Resizing and cropping

The classification model aims to identify the subject, while the segmentation model groups object per pixel. Hence, cropping and resizing are acceptable transformations at relatively higher safe parameters. The key code line for the wrapper function is as follows:

```
# use albumentations function for resizing and cropping
aug_album = albumentations.RandomSizedCrop(
  min_max_height=(500, 600),
  height=500,
  width=500,
  p=1.0)
```

The command for randomly resizing and cropping the CamVid segmentation data in the Python Notebook is as follows:

```
# use resize and crop wrapper functiion for camvid photo
pluto.draw_image_resize_segmen(pluto.df_camvid)
```

The output is as follows:

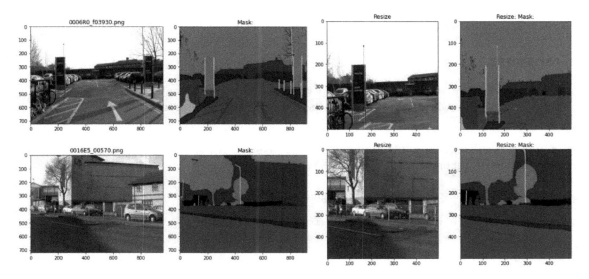

Figure 4.16 – Resizing and cropping the CamVid dataset

The command for randomly resizing and cropping the aerial segmentation data is as follows:

```
# use resize and crop wrapper functiion for aerialphoto
pluto.draw_image_resize_segmen(pluto.df_aerial)
```

The output is as follows:

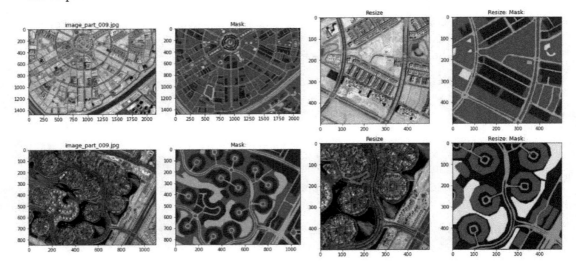

Figure 4.17 – Resizing and cropping the aerial dataset

Next, we'll cover the transpose filter.

Transpose

Pluto didn't use a transpose filter in *Chapter 3* for classification, but it is permissible for segmentation. Transposing involves switching the *x axis* with the *y axis*. The key code line for the wrapper function is as follows:

```
# use albumentations library for transpose
aug_album = albumentations.Transpose(p=1.0)
```

The command for transposing the CamVid segmentation data is as follows:

```
# use transpose wrapper function for camvid data
pluto.draw_image_transpose_segmen(pluto.df_camvid)
```

The output is as follows:

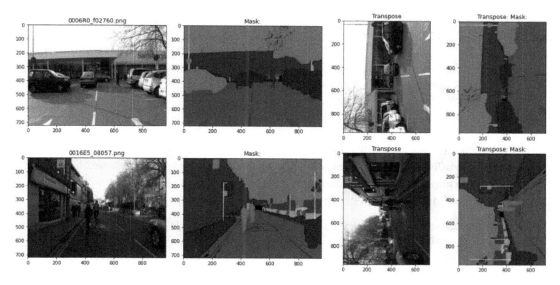

Figure 4.18 – Transposing the CamVid dataset

The command for transposing the aerial segmentation data is as follows:

```
# use transpose wrapper function for aerial data
pluto.draw_image_transpose_segmen(pluto.df_aerial)
```

The output is as follows:

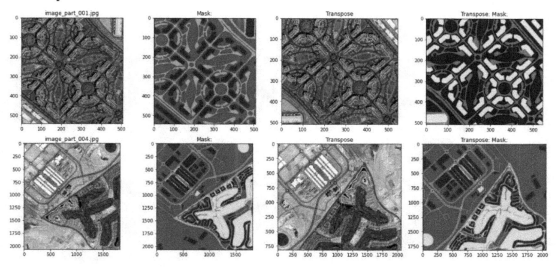

Figure 4.19 – Transposing the aerial dataset

> **Fun challenge**
>
> Implement optical distortion in the Python Notebook. Hint: use a similar Pluto wrapper function to the Albumentations library function's `albumentations.OpticalDistortion()` method.

Transpose is the last example Pluto uses for geometric transformations. Lighting, also known as brightness, belongs to the photometric transformations class.

Lighting

Lighting or brightness is acceptable for segmentation in the Albumentations library, but it belongs to the photometric transformations class. The original image changes to a random brightness level up to a safe level, but the mask image will not change. For both datasets, the safe parameter is a brightness of 0.5. The key code line in the wrapper function is as follows:

```
# use albumentations library for brightness
aug_album = albumentations.ColorJitter(
    brightness=brightness,
    contrast=0.0,
    saturation=0.0,
    hue=0.0,
    always_apply=True,
    p=1.0)
```

The command for using lighting on the CamVid segmentation data is as follows:

```
# use the brightmess wrapper function for camvid image
pluto.draw_image_brightness_segmen(pluto.df_camvid)
```

The output is as follows:

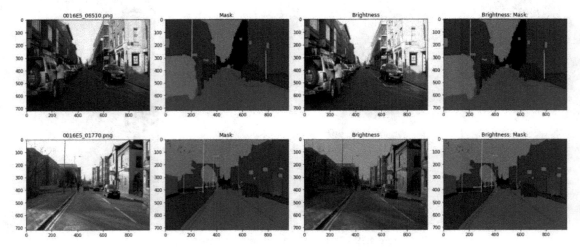

Figure 4.20 – Using lighting on the CamVid dataset

The command for using lighting on the aerial segmentation data is as follows:

```
# use the brightmess wrapper function for aerial image
pluto.draw_image_brightness_segmen(pluto.df_aerial)
```

The output is as follows:

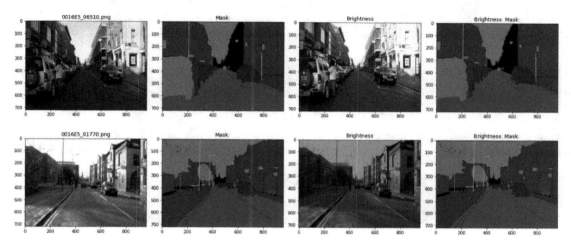

Figure 4.21 – Using lightning on the aerial dataset

Similar to the lighting filter, FancyPCA belongs to the photometric transformations class.

FancyPCA

FancyPCA is the last example Pluto demonstrates for photometric transformations. For both datasets, the safe parameter is an alpha value of 0.3. Once again, FancyPCA will not alter the mask image. The key code line in the wrapper function is as follows:

```
# use albumentations library for fancyPCA
aug_album = albumentations.FancyPCA(
  alpha=alpha,
  always_apply=True,
  p=1.0)
```

The command for using FancyPCA on the CamVid segmentation data is as follows:

```
# use the fancyPCA wrapper function for camvid image
pluto.draw_image_fancyPCA_segmen(pluto.df_camvid)
```

The output is as follows:

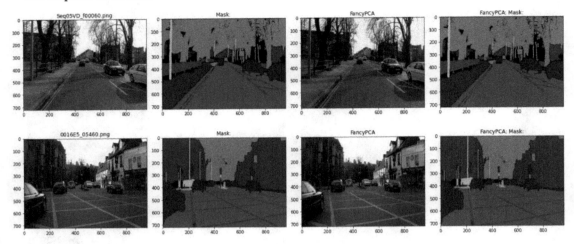

Figure 4.22 – Using FancyPCA on the CamVid dataset

The command for using FancyPCA on the aerial segmentation data is as follows:

```
# use the fancyPCA wrapper function for aerial image
pluto.draw_image_fancyPCA_segmen(pluto.df_aerial)
```

The output is as follows:

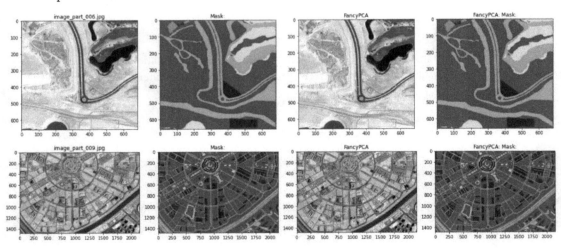

Figure 4.23 – Using FancyPCA on the aerial dataset

> **Fun challenge**
> Here is a thought experiment or maybe a practice one too: what can you do that appears acceptable in image augmentation but has a high probability of a **false-positive** or **false-negative** prediction in real-world deployment? Sorry, no hint.

Pluto finds that segmentation augmentation is not that different from classification augmentation. The wrapper functions are virtually the same, and only the helper methods display the images differently. Pluto has demonstrated segmentation for the flipping, resizing, cropping, rotating, transposing, lighting, and FancyPCA transformations. Similarly to *Chapter 3*, next, Pluto will combine individual filters into an uber function.

Combining

Before coding the uber combination methods in Python, Pluto needs to use pandas to summarize the filters in this chapter. Many more transformations are applicable for segmentation, so if you experiment with other filters in the Python Notebook, expand the pandas table with your new filters.

Pluto displays the summary table using the following command:

```
# use Pandas to display the combination filters
pluto.print_safe_parameters_segmen()
```

The output is as follows:

	filters	CamVid	CamVid Mask	Aerial	Aerial Mask
0	Horizontal Flip	Yes	Yes	Yes	Yes
1	Vertical Flip	Yes	Yes	Yes	Yes
2	Resize and Crop	Yes	Yes	Yes	Yes
3	Rotation	Yes	Yes	Yes	Yes
4	Transpose	Yes	Yes	Yes	Yes
5	Lighting	Yes	No	Yes	No
6	FancyPCA	Yes	No	Yes	No

Figure 4.24 – Summary segmentation filters

Using the summary table, Pluto writes the wrapper function. The key code line is as follows:

```
# use albumentations library
aug_album = albumentations.Compose([
```

```
albumentations.ColorJitter(brightness=0.5,
  contrast=0.0,
  saturation=0.0,
  hue=0.0,p=0.5),
albumentations.HorizontalFlip(p=0.5),
albumentations.Flip(p=0.5),
albumentations.Rotate(limit=45, p=0.5),
albumentations.RandomSizedCrop(
  min_max_height=(500, 600),
  height=500,
  width=500,
  p=0.5),
albumentations.Transpose(p=0.5),
albumentations.FancyPCA(alpha=0.2, p=0.5)])
```

Pluto displays the combination segmentation transformations for the CamVid dataset as follows:

```
# use combination wrapper function for camvid photo
pluto.draw_uber_segmen(pluto.df_camvid)
```

The output is as follows:

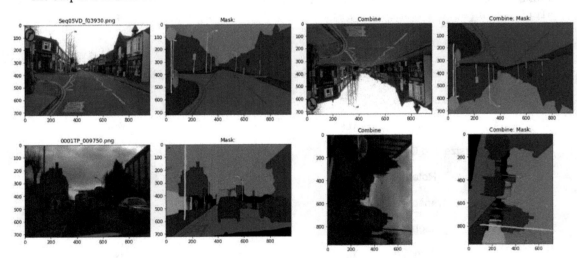

Figure 4.25 – Combining the filters for the CamVid dataset

The command for the aerial dataset in the Python Notebook is as follows:

```
# use combination wrapper function for aerial photo
pluto.draw_uber_segmen(pluto.df_aerial)
```

The output is as follows:

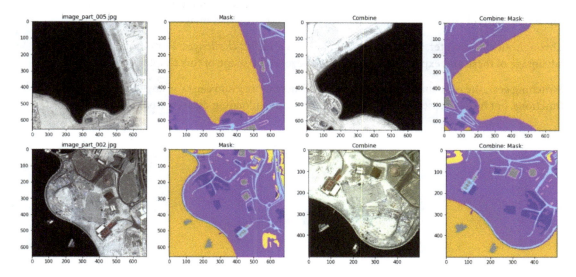

Figure 4.26 – Combining filters for the aerial dataset

> **Fun challenge**
>
> Pluto challenges you to refactor the **Pluto class** to make it faster and more compact. You are encouraged to create and upload your library to *GitHub and PyPI.org*. Furthermore, you don't have to name the class **PacktDataAug**, but it would give Pluto and his human companion a great big smile if you cited or mentioned this book. The code goals were for ease of understanding, reusable patterns, and teaching on the **–Python Notebook**. Thus, refactoring the code as a Python library would be relatively painless and fun.

With that, you've learned how to combine segmentation transformations. Next, we'll summarize what was covered in this chapter.

Summary

Image segmentation consists of the original image and an accompanying mask image. The goal is to determine whether a pixel belongs to a list of objects. For example, an urban photograph consists of streets, street signs, cars, trucks, bicycles, buildings, trees, and pedestrians. Image segmentation's job is to decide whether this pixel belongs to a car, tree, or other objects.

Image segmentation and image classification share the same transformations. In other words, most geometric transformations, such as flipping, rotating, resizing, cropping, and transposing, work with the original image and mask image in image segmentation. Photometric transformations, such as brightness, contrast, and FancyPCA, can technically be done with Python, but the filter does not alter the mask image. On the other hand, filters such as noise injection and random erasing are unsuitable for segmentation because they add or replace pixels in the original image.

Throughout this chapter, there have been *fun facts* and *fun challenges*. Pluto hopes you will take advantage of these and expand your experience beyond the scope of this chapter.

Switching gear, the next chapter will cover text augmentation. Pluto can't use any image augmentation functions, but he can reuse the wrapper functions for downloading datasets from the *Kaggle* website.

Part 3:
Text Augmentation

This part includes the following chapters:

- *Chapter 5, Text Augmentation*
- *Chapter 6, Text Augmentation with Machine Learning*

5
Text Augmentation

Text augmentation is a technique that is used in **Natural Language Processing** (**NLP**) to generate additional data by modifying or creating new text from existing text data. Text augmentation involves techniques such as character swapping, noise injection, synonym replacement, word deletion, word insertion, and word swapping. Image and text augmentation have the same goal. They strive to increase the size of the training dataset and improve AI prediction accuracy.

Text augmentation is relatively more challenging to evaluate because it is not as intuitive as image augmentation. The intent of an image augmentation technique is clear, such as flipping a photo, but a character-swapping technique will be disorienting to the reader. Therefore, readers might perceive the benefits as subjective.

The effectiveness of text augmentation depends on the quality of the generated data and the specific NLP task being performed. It can be challenging to determine the appropriate *safe* level of text augmentation that is required for a given dataset, and it often requires experimentation and testing to achieve the desired results.

Customer feedback or social media chatter is fair game for text augmentation because the writing is messy and, predominantly, contains grammatical errors. Conversely, legal documents or written medical communications, such as doctor's prescriptions or reports, are off-limits because the message is precise. In other words, error injections, synonyms, or even AI-generated text might change the legal or medical meaning beyond the *safe* level.

The biases in text augmentation are equally difficult to discern. For example, adding noise by purposely misspelling words using the Keyboard augmentation method might introduce bias against real-world tweets, which typically contain misspelled words. There are no generalized rules to follow, and the answer only becomes evident after thoroughly studying the data and reviewing the AI forecasting objective.

> **Fun fact**
>
> As generative AI becomes more widely available, you can use **OpenAI's GPT-3**, **Google Bard**, or **Facebook's Roberta** system to generate original articles for text augmentation. For example, you can ask generative AI to create positive or negative reviews about a company product, then use the AI-written articles to train predictive AI on sentiment analysis.

In *Chapter 5*, you will learn about text augmentation and how to code the methods in Python notebooks. In particular, we will cover the following topics:

- Character augmenting
- Word augmenting
- Sentence and flow augmenting
- Text augmentation libraries
- Real-world text datasets
- Reinforcing learning through Python Notebook

Let's get started with the simplest topic, character augmentation.

Character augmenting

Character augmentation substitutes or injects characters into the text. In other words, it creates typing errors. Therefore, the method seems counterintuitive. Still, just like noise injection in image augmentation, scholarly published papers illustrate the benefit of character augmentation in improving AI forecasting accuracy, such as *Effective Character-Augmented Word Embedding for Machine Reading Comprehension* by *Zhuosheng Zhang, Yafang Huang, Pengfei Zhu, and Hai Zhao*, from the 2018 *CCF International Conference on Natural Language Processing*.

The three standard methods for character augmentation are listed as follows:

- The **Optical Character Recognition (OCR) augmenting** function substitutes frequent errors in OCR by converting images into text, such as the letter *o* into the number *0* or the capital letter *I* into the number *1*.

- The **Keyboard augmenting** method replaces a character with other characters that are adjacent to it. For example, a typical typing error for character *b* is hitting key *v* or key *n* instead.

- The **Random character** function randomly swaps, inserts, or deletes characters within the piece of text.

> **Fun fact**
>
> Computer encoding text was very different from 1963 to 1970; for example, a computer would encode the letter *A* as an integer 64 or Hexidecimal 41. This originated from the **American National Standards Institute (ANSI)** in 1964, and the **International Organization for Standardization (ISO)** adopted the standard around 1970. In 1980, the **Unification Code (Unicode)** subsumed the ISO standard for all international languages. However, if you come across computer text from around 1964, it could be encoded in **Extended Binary Coded Decimal Interchange Code (EBCDIC)**, which encodes the letter *A* as 193 or Hexidecimal C1. As a programmer, you might have to answer this question: *does your website or mobile app support Unicode?*

After character augmenting, the next category is word augmenting.

Word augmenting

Word augmentations carry the same bias and *safe* level warning as character augmentations. Over half of these augmentation methods inject errors into the text, but other functions generate new text using synonyms or a pretrained AI model. The standard word augmentation functions are listed as follows:

- The **Misspell augmentation** function uses a predefined dictionary to simulate spelling mistakes. It is based on the scholarly paper *Text Data Augmentation Made Simple By Leveraging NLP Cloud APIs* by **Claude Coulombe**, which was published in 2018.

- The **Split augmentation** function splits words into two tokens randomly.

- The **Random word** augmentation method applies random behavior to the text with four parameters: **substitute**, **swap**, **delete**, and **crop**. It is based on two scholarly papers: *Synthetic and Natural Noise Both Break Neural Machine Translation* by **Yonatan Belinkov and Yonatan Bisk**, published in 2018, and *Data Augmentation via Dependency Tree Morphing for Low-Resource Languages* by **Gozde Gul Sahin and Mark Steedman**.

- The **Synonym augmentation** function substitutes words with synonyms from a predefined database. The first option is to use **WordNet**. WordNet is an extensive lexical database of English from *Princeton University*. The database groups nouns, verbs, adjectives, and adverbs into sets of cognitive synonyms. The second option is to use a **Paraphrase Database (PPDB)**. A PPDB is an automatically extracted database containing millions of paraphrases in 16 languages. A PPDB aims to improve language processing by making systems more robust to language variability and unseen words. The entire PPDB resource is freely available under the United States **Creative Commons Attribution 3.0** license.

- The **Antonym augmentation** function replaces words with antonyms. It is based on the scholarly paper, *Adversarial Over-Sensitivity and Over-Stability Strategies for Dialogue Models*, by **Tong Niu and Mohit Bansal**, which was published in 2018.

- The **Reserved Word augmentation** method swaps target words where you define a word list. It is the same as synonyms, except the terms are created manually.

> **Fun challenge**
> Here is a thought experiment: can you think of a new character or word augmentation technique? A hint is to think about how a dyslexic person reads.

Next, we will look at sentence augmentation.

Sentence augmentation

Sentence augmenting uses generative AI to create new texts. Examples of AI models are BERT, Roberta, GPT-2, and others.

The three sentence augmentation methods are listed as follows:

- **Contextual Word Embeddings** uses GPT-2, Distilled-GPT-2, and XLNet.
- **Abstractive Summarization** uses Facebook Robertaand T5-Large.
- **Top-n Similar Word** uses LAMBADA.

Before Pluto explains the code in the Python Notebook, let's review the text augmentation libraries.

Text augmentation libraries

There are many more Python open source image augmentation libraries than text augmentation libraries. Some libraries are more adaptable to a particular category than others, but in general, it is a good idea to pick one or two and become proficient in them.

The well-known libraries are **Nlpaug**, **Natural Language Toolkit (NLTK)**, **Generate Similar (Gensim)**, **TextBlob**, **TextAugment**, and **AugLy**:

- **Nlpaug** is a library used for textual augmentation for DL. The goal is to improve DL model performance by generating textual data. The GitHub link is `https://github.com/makcedward/nlpaug`.

- **NLTK** is a platform used for building Python programs to work with human language data. It provides interfaces to over 50 corpora and lexical resources, such as WordNet. NLTK contains text-processing libraries for classification, tokenization, stemming, tagging, parsing, and semantic reasoning. The GitHub link is `https://github.com/nltk/nltk`.

- **Gensim** is a popular open source NLP library used for unsupervised topic modeling. It uses academic models and modern statistical machine learning to perform word vectors, corpora, topic identification, document comparison, and analyzing plain-text documents. The GitHub link is `https://github.com/RaRe-Technologies/gensim`.

- **TextBlob** is a library that is used for processing textual data. It provides a simple API for diving into typical NLP tasks such as part-of-speech tagging, noun phrase extraction, sentiment analysis, classification, and translation. The GitHub link is `https://github.com/sloria/TextBlob`.

- **TextAugment** is a library that is used for augmenting text in NLP applications. It uses and combines the NLTK, Gensim, and TextBlob libraries. The GitHub link is `https://github.com/dsfsi/textaugment`.

- **AugLy** is a data augmentation library from Facebook that supports audio, image, text, and video modules and over 100 augmentations. The augmentation of each modality is categorized into sub-libraries. The GitHub link is `https://github.com/facebookresearch/AugLy`

Similarly to image augmentation wrapper functions, Pluto will write wrapper functions that use the library under the hood. You can pick more than one library for a project, but Pluto will use the **Nlpaug** library to power the wrapper functions.

Let's start by downloading the real-world text datasets from the *Kaggle* website.

Real-world text datasets

The *Kaggle* website is an online community platform for data scientists and machine learning enthusiasts. The Kaggle website has thousands of real-world datasets; Pluto found a little over 2,900 **NLP** datasets and has selected two NLP datasets for this chapter.

In *Chapter 2*, Pluto uses the **Netflix** and **Amazon** datasets as examples with which to understand biases. Pluto keeps the **Netflix** NLP dataset because the movie reviews are curated . There are a few syntactical errors, but overall, the input texts are of high quality.

The second **NLP** dataset is **Twitter Sentiment Analysis (TSA)**. The 29,530 real-world tweets contain many grammatical errors and misspelled words. The challenge is to classify the tweets into two categories: (1) normal or (2) racist and sexist.

The dataset was published in 2021 by **Mayur Dalvi**, and the license is **CC0: Public Domain**, `https://creativecommons.org/publicdomain/zero/1.0/`.

After selecting the two NLP datasets, you can use the same four steps to begin the process of practical learning through a Python Notebook. If you need clarification, review *Chapters 2* and *3*. The steps are listed as follows:

1. Retrieve Python Notebook and Pluto.
2. Download real-world data.
3. Import into pandas.
4. View data.

Let's start with Pluto.

The Python Notebook and Pluto

Start by loading the `data_augmentation_with_python_chapter_5.ipynb` file into **Google Colab** or your chosen Jupyter notebook or JupyterLab environment. From this point onward, the code snippets are from the Python Notebook, which contains the complete code.

The next step is to clone the repository. Pluto will reuse the code from *Chapter 2* because it has the downloading Kaggle data methods and not the image augmentation functions. The `!git` and `%run` statements are used to start up Pluto. The command is as follows:

```
# clone GitHub repo
!git clone 'https://github.com/PacktPublishing/Data-Augmentation-with-Python'
# instantiate Pluto
%run 'Data-Augmentation-with-Python/pluto/pluto_chapter_2.py'
```

The output is as follows:

```
--------------------------- : ---------------------------
          Hello from class : <class '__main__.PacktDataAug'> Class:
PacktDataAug
                Code name : Pluto
                Author is : Duc Haba
--------------------------- : ---------------------------
```

We need one more check to ensure Pluto has been loaded satisfactorily. The following command asks Pluto to say his status:

```
pluto.say_sys_info()
```

The output should be as follows or similar, depending on your system:

```
--------------------------- : ---------------------------
              System time : 2022/10/30 06:52
                 Platform : linux
    Pluto Version (Chapter) : 2.0
          Python (3.7.10) : actual: 3.7.15 (default, Oct 12 2022,
19:14:55) [GCC 7.5.0]
          PyTorch (1.11.0) : actual: 1.12.1+cu113
           Pandas (1.3.5) : actual: 1.3.5
              PIL (9.0.0) : actual: 7.1.2
       Matplotlib (3.2.2) : actual: 3.2.2
                CPU count : 2
               *CPU speed : NOT available
--------------------------- : ---------------------------
```

Here, Pluto reported that he is from *Chapter 2*, which is also known as **version 2.0**. This is what we wanted because we don't need any image augmentation functions from *Chapters 3* and *4*. The next step is to download the real-world **Netflix** and **Twitter** NLP datasets.

Real-world NLP datasets

There has yet to be any new code written for this chapter. Pluto reuses the `fetch_kaggle_dataset()` method to download the **Netflix** dataset, as follows:

```
# fetch data
url = 'https://www.kaggle.com/datasets/infamouscoder/dataset-netflix-shows'
pluto.fetch_kaggle_dataset(url)
```

The `dataset-netflix-shows.zip` file is 1.34 MB, and the function automatically unzips in the **kaggle** directory.

The method for fetching the Twitter dataset is as follows:

```
# fetch data
url = 'https://www.kaggle.com/datasets/mayurdalvi/twitter-sentiments-analysis-nlp'
pluto.fetch_kaggle_dataset(url)
```

The `twitter-sentiments-analysis-nlp.zip` file is 1.23 MB, and the function automatically unzips in the **kaggle** directory.

> **Fun challenge**
>
> The challenge is to search and download two additional real-world NLP datasets from the Kaggle website. Hint: use the `pluto.fetch_kaggle_dataset()` method. Pluto is an imaginary digital Siberian Husky. Therefore, he will happily fetch data until your disk space is full.

The next step is to load the data into pandas.

Pandas

Pandas is the de facto standard for data scientists to manage and manipulate tabular data. It is fast, flexible, easy to use, and powerful. Therefore, Pluto uses pandas to import the **Comma-Separated Values (CSV)** file, and he reuses the `fetch_df()` method. Note that `df` is a typical shorthand for the pandas **DataFrame** class.

For the **Netflix** data, Pluto uses the following two commands for importing to pandas and printing out the data batch:

```
# import into Panda
f = 'kaggle/dataset-netflix-shows/netflix_titles.csv'
pluto.df_netflix_data = pluto.fetch_df(f)
# display data batch
pluto.print_batch_text(pluto.df_netflix_data,
    cols=['title', 'description'])
```

The output is as follows:

	title	description
391	Wynonna Earp	The outcast descendant of lawman Wyatt Earp teams up with an immortal Doc Holliday to rid the world of demonic revenants from the Wild West.
7629	Nurses Who Kill	Top medical, criminal and psychological experts analyze the motives and methods of nurses who use their positions to kill rather than heal.
8391	The Legend of Michael Mishra	After a life of crime, a notorious kidnapper tries to change his ways and turn over a new leaf to win the heart of the woman he loves.
2157	Wizards: Tales of Arcadia	Merlin's apprentice joins Arcadia's heroes on a time-bending adventure in Camelot, where conflict is brewing between the human, troll and magical worlds.
1179	Mighty Morphin Power Rangers	Five average teens are chosen by an intergalactic wizard to become the Power Rangers, who must use their new powers to fight the evil Rita Repulsa.
2526	She-Ra and the Princesses of Power	Soldier Adora finds a magic sword and her identity as legendary hero She-Ra. She joins the Rebellion, but her best friend stays with the evil Horde.

Figure 5.1 – Netflix movie descriptions

> **Fun fact**
>
> The `fetch_df()` method randomly selects several records to display in the data batch. The number of records, or batch size, is the `bsize` parameter. The default is 10 records.

The **Netflix** movie-reviewed data is curated; therefore, it is clean. Pluto doesn't have to scrub the data. However, the **Twitter** data is another story.

The commands for cleaning, importing, and batch-displaying the **Twitter** data to pandas are as follows:

```
# clean space-char
f = 'kaggle/twitter-sentiments-analysis-nlp'
!find {f} -name "* *" -type f | rename 's/ /_/g'
# import into Pandas
f = 'kaggle/twitter-sentiments-analysis-nlp/Twitter_Sentiments.csv'
pluto.df_twitter_data = pluto.fetch_df(f)
# display data batch
pluto.print_batch_text(pluto.df_twitter_data,cols=['label', 'tweet'])
```

The output is as follows:

	label	tweet
29602	1	#southsudan provoking neighbours again with noise. the #scumbag that says - he can do what ever he wants. #uncivilised trash. haters
28346	0	glad it worked well for you!!! please let me know what you use and if i can help!
17440	0	so ð !!! going to have a show with @user and jessi this sunday! @user 3 girls in 1 hotel room. that sounds fun ð â ¤
15164	0	too sensitive or do fans have a right to be with #xmenapocalypse movie poster?
11153	0	@user he was hit on foot so until the cars around us chill we could all end up like him
8613	0	then it dawned on me that he's not my lil baby brother and eventually he'll be leaving for college.

Figure 5.2 – Twitter tweets

Since the real-world tweets from **Twitter** have been written by the public, they contain misspelled words, bad words, and all sorts of shenanigans.

The goal is to predict regular versus racist or sexist tweets. Pluto focuses on learning text argumentation; therefore, he prefers to have tweets with printable characters, no HTML tags, and no words of profanity.

Pluto writes two simple helper methods to clean the text and remove the words of profanity. The _clean_text() function uses the regex library, and the one line of code is as follows:

```
return (re.sub('[^A-Za-z0-9 .,!?#@]+', '', str(x)))
```

The _clean_bad_word() helper function uses the filter-profanity library, and the one line of code is as follows:

```
return (profanity.censor_profanity(x, ''))
```

The `clean_text()` method uses the two helper functions with pandas' powerful `apply` function. Using pandas' built-in functions, Pluto writes the `clean_text()` function with two code lines instead of a dozen lines using standard `if-else` and `for`-loop construct. The code is as follows:

```
# clean text
df['clean_tweet'] = df.tweet.apply(self._clean_text)
# remove profanity words
df['clean_tweet'] = df['clean_tweet'].apply(
    self._clean_bad_word)
```

The commands for the clean tweets and showing the data batch are as follows:

```
# clean tweets
pluto.clean_text(pluto.df_twitter_data)
# display data batch
pluto.print_batch_text(pluto.df_twitter_data,
    cols=['label', 'clean_tweet'])
```

The output is as follows:

	label	clean_tweet
24436	0	@user my amp go out 2the victims ampfamilies of this horrible attack in orlando. if ur a @user voter think twice #sa
25183	0	@user thanks unc @user ...had #fun #shopping #today w my #uncle #vans #converse
2475	1	@user if you are #hispanic #black #asian a nonracist white, other minority dont serve in military while in power #cnn
12228	0	@user whos excited for robinwood activitiy weekend? see you all tomorrow #early night @user
4342	0	@user too excited to see @user tomorrow!!!!!!! with all my #positiveposse @user @user @user #the
27026	1	@user last retro #christmasadve for you this year, and whilst i do love a good bit of #tupperwear i do not want

Figure 5.3 – Clean Twitter tweets

> **Fun fact**
>
> Who would have known that a dog and a panda could work together well? The next *Kung Fu Panda* movie is about **Po** and **Pluto** teaming up to defend and augment the city wall against the storm of the century, which has been caused by global warming.

Let's use pandas and some other libraries to visualize the NLP dataset.

Visualizing NLP data

Chapter 2 uses the `draw_word_count()` method to display the average word per record and the shortest and longest movie reviews. The right-hand side of the graph shows the histogram of the movie review word counts. The pandas library generates beautiful word count charts. Pluto reuses the function to display the **Netflix** NLP data, as follows:

```
# draw word count
pluto.draw_word_count(pluto.df_netflix_data)
```

The output is as follows:

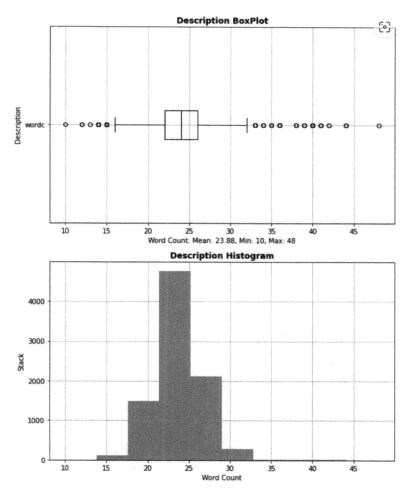

Figure 5.4 – Netflix word counts

The Netflix movie description mean is 23.88 words, with a minimum of 10 words and a maximum of 48 words. Pluto does the same for the **Twitter** NLP data, as follows:

```
# draw word count
pluto.draw_word_count(pluto.df_twitter_data)
```

The output is as follows:

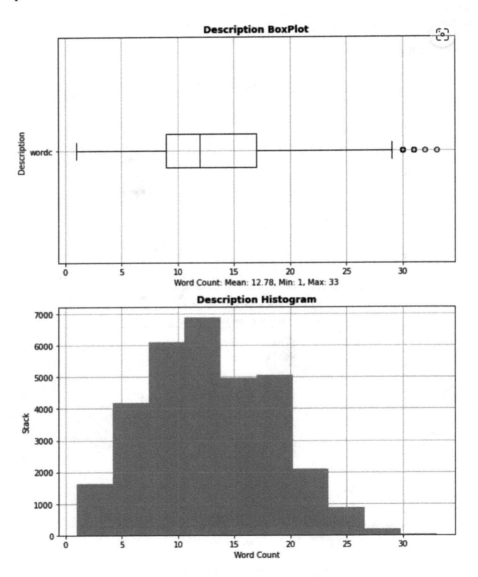

Figure 5.5 – Twitter word counts

The average word count of the Twitter tweets is 12.78 words, with a minimum of 1 word and a maximum of 33 words.

Pluto writes the `draw_text_null_data()` method to check whether there is any missing data, also known as a **null** value. The missing data shows up as a white line. The `Missingno` library generates the graph with the following key line of code:

```
missingno.matrix(df,color=color,ax=pic)
```

Pluto draws the `null` data graph for the Netflix data, as follows:

```
# draw missing data/null value
pluto.draw_text_null_data(pluto.df_netflix_data)
```

The output is as follows:

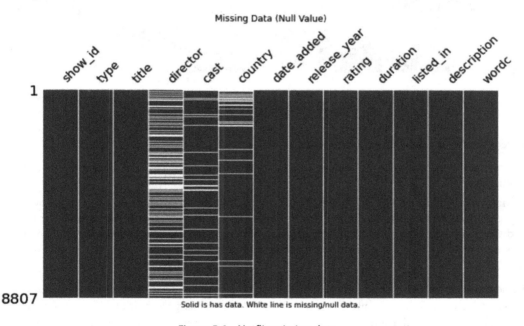

Figure 5.6 – Netflix missing data

There is missing data in the **director**, **cast**, and **country** categories for the **Netflix** data, but the **description** category, also known as the movie review, has no missing data.

Pluto does the same for the **Twitter** data, as follows:

```
# draw missing data/null value
pluto.draw_text_null_data(pluto.df_twitter_data)
```

The output is as follows:

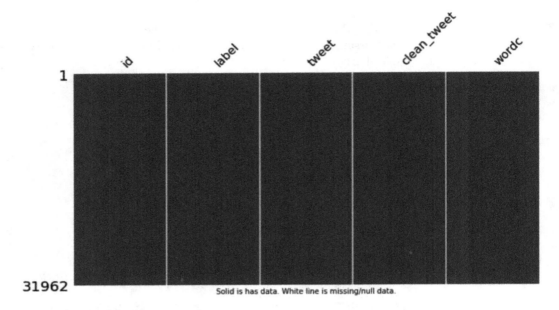

Figure 5.7 – Twitter missing data

There is no missing data in the **Twitter** data.

> **Fun fact**
>
> Many multimillion-dollar AI systems have failed primarily because of a lack of control over the input data. For example, the *Amazon Recruiting* system in 2020 failed because there was no diversity in the dataset, and the most egregious debacle was *Microsoft's Chatbot Tay* in 2016. It was corrupted by Twitter users inputting sexist and racist tweets.

The next chart is the word cloud infographic diagram. This is an extraordinary method for visualizing the NLP text. The most commonly used words are displayed in a large font, while the least used terms are displayed in a smaller font. The **WordCloud** library generates the infographic chart, and the essential code snippet is as follows:

```
# generate word cloud
img = wordcloud.WordCloud(width = 1600,
        height = 800,
        background_color ='white',
        stopwords = xignore_words,
        min_font_size = 10).generate(words_str)
# covert Pandas to word string input
```

```
orig = df_1column.str.cat()
word_clean = re.sub('[^A-Za-z0-9 ]+', '', orig)
```

Pluto uses the _draw_text_wordcloud() helper function and the draw_text_wordcloud() method to display the infographic chart for real-world **Netflix** data, as follows:

```
# draw word cloud
pluto.draw_text_wordcloud(pluto.df_netflix_data.description,
    xignore_words=wordcloud.STOPWORDS,
    title='Word Cloud: Netflix Movie Review')
```

The output is as follows:

Figure 5.8 – Netflix word cloud, with approximately 246,819 words

Pluto does the same for the real-world **Twitter** data, as follows:

```
# draw word cloud
pluto.draw_text_wordcloud(pluto.df_twitter_data.clean_tweet,
    xignore_words=wordcloud.STOPWORDS,
    title='Clean Tweets Word Cloud')
```

The output is as follows:

Figure 5.9 – Twitter   word cloud, with approximately 464,992 words

---

Fun fact

Here is a fun fact about the history of word cloud graphs. The word cloud, also known as a tag cloud, Wordle, or weighted list, was first used in print by **Douglas Coupland** in the book *Microserfs*. It was published in 1995, but not until 2004 did the word clouds exist in digital format on the *Flickr* website. Today, word cloud infographics are widespread on the web and in academic papers.

So far, Pluto has discussed character, word, and sentence augmentation theories, chosen the **Nlpaug** text augmentation library, and downloaded the real-world **Netflix** and **Twitter** NLP datasets. It is time for Pluto to reinforce his learning by performing text augmentation with Python code.

Reinforcing learning through Python Notebook

Pluto uses the Python Notebook to reinforce our understanding of text augmentation. He uses the batch function to display text in batches. This works similarly to the batch functions for images. In other words, it randomly selects new records and transforms them using the augmentation methods.

> **Fun fact**
>
> Pluto recommends running the batch functions repeatedly to gain a deeper insight into the text augmentation methods. There are thousands of text records in the **Twitter** and **Amazon** datasets. Each time you run the batch functions, it displays different records from the dataset.

As with the image augmentation implementation, the wrapper functions use the **Nlpaug** library under the hood. The wrapper function allows you to focus on the text transformation concepts and not be distracted by the library implementation. You can use another text augmentation library, and the wrapper function input and output will remain the same.

Pluto could write one complex function that contains all the text transformation techniques, and it may be more efficient, but that is not the goal of this book. After reading this book, you can choose to rewrite or hack the Python Notebook to suit your style with confidence.

In this chapter, Pluto uses an opening line from the book *A Tale of Two Cities* by **Charles Dickens** as the control text. Pluto paraphrases the text by substituting the commas between the phrases with periods because this makes it easier for the text augmentation process. The control text is as follows:

"It was the best of times. It was the worst of times. It was the age of wisdom. It was the age of foolishness. It was the epoch of belief. It was the epoch of incredulity."

The Python Notebook covers the following topics:

- Character augmentation
- Word augmentation

Let's start with the three character augmentation techniques.

Character augmentation

Character augmentation involves injecting errors into the text. The process is counterintuitive because it purposely adds errors to the data. In other words, it makes the text harder for humans to understand. In contrast, computers use deep learning algorithms to predict the outcome, particularly the **Convolutional Neural Network** (**CNN**) and the **Recurrent Neural Network** (**RNN**) algorithms. For example, sentiment classification for tweets does not affect by misspelled words.

In particular, Pluto will explain the following three methods:

- OCR augmenting
- Keyboard augmenting
- Random augmenting

Let's start with OCR.

OCR augmenting

The OCR process converts an image into a piece of text, with frequent errors such as mixing *0* and *o* during the conversion.

Pluto writes the `_print_aug_batch()` helper function to randomly select sample records from the NLP data, apply the text augmenting method, and print it using pandas. The input or method definition is as follows:

```
# method definition
def _print_aug_batch(self, df,
    aug_func,
    col_dest="description",
    bsize=3,
    aug_name='Augmented'):
```

Here, `df` is the pandas DataFrame, `aug_function` is the augmentation method from the wrapper function, `col_dest` is the chosen column destination, `bsize` is the number of samples in the batch with a default of three, and `title` is the optional title for the chart.

The OCR wrapper function is elementary. The two lines of code are the **Nlpaug** library text augmentation method (`aug_func`) and the helper function. The entire code is as follows:

```
# method definiton
@add_method(PacktDataAug)
def print_aug_ocr(self, df,
    col_dest="description",
    bsize=3,
    aug_name='Augmented'):
    aug_func = nlpaug.augmenter.char.OcrAug()
    self._print_aug_batch(df,
        aug_func,
        col_dest=col_dest,
        bsize=bsize,
        aug_name=aug_name)
    return
```

Pluto uses the `print_aug_ocr()` method with the **Netflix** data, as follows:

```
# use OCR method
pluto.print_aug_ocr(pluto.df_netflix_data,
    col_dest='description',
    aug_name='OCR Augment')
```

The output is as follows:

	OCR Augment	Original
0	1t was the 6e8t of times. It was the worst of times. 1t was the a9e of wisdom. It was the age of fu0lishne88. It was the epoch of 6e1ief. 1t was the epoch of incredulity.	It was the best of times. It was the worst of times. It was the age of wisdom. It was the age of foolishness. It was the epoch of belief. It was the epoch of incredulity.
1	After experiencing a tka9ic loss, a woman must re8i8t a new fami1y dynamic that could control the fotuke of her fathers' company and her 1ife.	After experiencing a tragic loss, a woman must resist a new family dynamic that could control the future of her fathers' company and her life.
2	The Kuryu Group mare8 it their mi88i0n to takeover the 8WORO district once and f0k all, but the 8tkeet 9an9 alliance has a plan 0f their uwn.	The Kuryu Group makes it their mission to takeover the SWORD district once and for all, but the street gang alliance has a plan of their own.
3	Two hapless cops find themselves in 0vek theik head8 as they cr08s path8 with dangerous criminals whi1e searching f0k Celestina, a 6e1uved goat mascot.	Two hapless cops find themselves in over their heads as they cross paths with dangerous criminals while searching for Celestina, a beloved goat mascot.

Figure 5.10 – Netflix OCR augmenting

In *Figure 5.10*, the first line is **Dickens'** control text, with the augmented text on the left-hand side and the original text on the right-hand side. The following three rows are randomly sampled from the **Netflix** NLP data. Pluto recommends that you read the left-hand augmented text first. Stop and try to decipher the meaning before reading the original text.

> **Fun fact**
>
> Pluto recommends repeatedly running the `print_aug_ocr()` method to see other movie descriptions. You can increase `bsize` to see more than two records at a time.

Pluto does the same for the **Twitter** NLP data, as follows:

```
# print the batch
pluto.print_aug_ocr(pluto.df_twitter_data,
    col_dest='clean_tweet',
    aug_name='OCR Augment')
```

The output is as follows:

	OCR Augment	Original
0	It was the best of times. It was the worst of times. 1t was the age uf wisdom. It wa8 the age 0f foolishness. It was the epoch 0f belief. 1t was the epoch of incredulity.	It was the best of times. It was the worst of times. It was the age of wisdom. It was the age of foolishness. It was the epoch of belief. It was the epoch of incredulity.
1	dk. frances cress we18in9 racism defined # whitesupremacy # drfranciscresswelsing	dr. frances cress welsing racism defined #whitesupremacy #drfranciscresswelsing
2	if we pentecostals weke as passionate about lu8t souls a8 we ake about politics wed turn the w0kld upside down.	if we pentecostals were as passionate about lost souls as we are about politics, we'd turn the world upside down.
3	jo8t want to sta my new job like nuw p18 # cantreepwaitin9 # newchapter	just want to sta my new job like now pls #cantkeepwaiting #newchapter

Figure 5.11 – Twitter OCR augmenting

Next, Pluto moves on from the OCR method to the keyboard technique.

Keyboard augmenting

The keyboard augmenting method replaces a character with a close-distance key on a keyboard. For example, a typical typing error for character *b* is using key *v* or key *n*.   The augmentation variable defines as follows:

```
# define augmentation function variable definition
aug_func = nlpaug.augmenter.char.KeyboardAug()
```

Pluto uses the print_aug_keyboard() wrapper function with the **Netflix** NLP data, as follows:

```
# use keyboard augmentation technique
pluto.print_aug_keyboard(pluto.df_netflix_data,
    col_dest='description',
    aug_name='Keyboard Augment')
```

The output is as follows:

	Keyboard Augment	Original
0	It was the beQ5 of timRX. It was the w8rs6 of tJmWs. It was the age of siCdom. It was the age of Eoo>iDhnesW. It was the @(och of b@kief. It was the 2poVh of LgcredklitT.	It was the best of times. It was the worst of times. It was the age of wisdom. It was the age of foolishness. It was the epoch of belief. It was the epoch of incredulity.
1	CoN3Rian Ben GI$ub TostZ thOz fSme Ehkw in which teams of two compete to solve increasingly tojTh gdain teasers. Faster answers mean more money.	Comedian Ben Gleib hosts this game show in which teams of two compete to solve increasingly tough brain teasers. Faster answers mean more money.
2	TNos drama based on reallife events tells the xYory of George MDKejja, the rouBh, dRt#emined new primcipX; of a noR(rKous Los Ange:wW high cchoop.	This drama based on reallife events tells the story of George McKenna, the tough, determined new principal of a notorious Los Angeles high school.
3	Tylers ztUll feeling lost fPll(wkng the tragedy thats syra9nrd his relationship with his father wh@j he mretc S?ly, a Hurl who Hnsers5anRs his pain.	Tylers still feeling lost following the tragedy thats strained his relationship with his father when he meets Ally, a girl who understands his pain.

Figure 5.12 – Netflix keyboard augmenting

Pluto does the same for the **Twitter** NLP data, as follows:

```
# use keyboard augmentation technique
pluto.print_aug_keyboard(pluto.df_twitter_data,
    col_dest='clean_tweet',
    aug_name='Keyboard Augment')
```

The output is as follows:

	Keyboard Augment	Original
0	It was the bfzt of ^Jmes. It was the worE4 of tim@a. It was the age of w7sdoh. It was the age of toolOshJeZs. It was the eLosh of belJ2f. It was the SpPch of &nceeculiyy.	It was the best of times. It was the worst of times. It was the age of wisdom. It was the age of foolishness. It was the epoch of belief. It was the epoch of incredulity.
1	wik: never understand people who dump fheif kiRA, damage that does cant b ciDed, i 0nlj holF 2 l)be them enough 2 help th@j 4Yivr # dbd	will never understand people who dump their kids, damage that does can't b fixed, i only hope 2 love them enough 2 help them 4give #dbd
2	can wh8teE only immigration make qme#ida grFAt ZRain? via @ TsDr # jarodtaylor # asslicker	can whites-only immigration make america great again? via @user #jarodtaylor #asslicker
3	mldjing! !! woke up VDeligg gd2at. toRZy. thW5e is # sHnsGin2. # Hon	morning!!! woke up feeling great. today. there is #sunshine . #nonke # #love

Figure 5.13 – Twitter keyboard augmenting

The last of the three text augmentation methods is the random technique.

Random augmenting

The random character function randomly swaps, inserts, or deletes characters in the text. The four modes for the random process are **inserting**, **deleting**, **substituting**, and **swapping**. The augmentation variable defines as follows:

```
# define augmentation function variable definition
aug_func = nlpaug.augmenter.char.RandomCharAug(action=action)
```

Pluto uses the `print_aug_random()` wrapper function with `action` set to `insert` in the **Netflix** NLP data, as follows:

```
# use random insert augmentation technique
pluto.print_aug_char_random(pluto.df_netflix_data,
    action='insert',
    col_dest='description',
    aug_name='Random Insert Augment')
```

The output is as follows:

	Random Insert Augment	Original
0	It was the ab+est of t6iMmes. It was the 7worsgt of t(imezs. It was the age of wisdqoZm. It was the age of bfaoAolis6hness. It was the ecpocFh of bfelilef. It was the epoZc)h of incureUdu4l+ity.	It was the best of times. It was the worst of times. It was the age of wisdom. It was the age of foolishness. It was the epoch of belief. It was the epoch of incredulity.
1	SuZddeinlly possessed woitBh supernatural kpoZwers, a fa$theJr sets out to help his esEtrdaqnged daughter, whos at #r!isk of losing evUeNrythikng sh)eEs lived for.	Suddenly possessed with supernatural powers, a father sets out to help his estranged daughter, who's at risk of losing everything she's lived for.
2	In a drama bkas_ed on a tVrHue sXto(ry, Bra4zi$liajn 4backpafckBer Gabriel Buchmann spejndws his gap yeea)r exhpldorinJg Africa. Why does he die in the MaOlaOwi mountains?	In a drama based on a true story, Brazilian backpacker Gabriel Buchmann spends his gap year exploring Africa. Why does he die in the Malawi mountains?
3	Fresh Army rreicrEuit Amy Cole is assigned to G3uUant_anamo Byayns CDaUmp MXRUay, w3hesre she fo^rJms a surprising boWnSd wUitph one of the prisoners in her charge.	Fresh Army recruit Amy Cole is assigned to Guantanamo Bays Camp XRay, where she forms a surprising bond with one of the prisoners in her charge.

Figure 5.14 – Netflix random insert augmenting

Pluto does the same for the **Twitter** NLP data, as follows:

```
# use random insert augmentation technique
pluto.print_aug_char_random(pluto.df_twitter_data,
    action='insert',
    col_dest='clean_tweet',
    aug_name='Random Insert Augment')
```

The output is as follows:

	Random Delete Augment	Original
0	It was the es of mes. It was the rst of ime. It was the age of wsdo. It was the age of lishnss. It was the eoh of blif. It was the poc of incrdli.	It was the best of times. It was the worst of times. It was the age of wisdom. It was the age of foolishness. It was the epoch of belief. It was the epoch of incredulity.
1	el said	well said
2	@ ur went back to the sre, on to in out th one of the young empoys had resigned, due to srespetfl tretmt	@user went back to the store, only to find out that one of the young employees had resigned, due to disrespectful treatment
3	i am alue. # iam # osive # ffirion	i am valued. #iam #positive #affirmation

Figure 5.15 – Twitter random insert augmenting

Pluto uses the `print_aug_random()` wrapper function with `action` set to `delete` for the **Netflix** NLP data, as follows:

```
# use random delete augmentation technique
pluto.print_aug_char_random(pluto.df_netflix_data,
    action='delete',
    col_dest='description',
    aug_name='Random Delete Augment')
```

The output is as follows:

	Random Delete Augment	Original
0	It was the bs of tms. It was the wos of ies. It was the age of wisd. It was the age of olisess. It was the eoc of beli. It was the poh of incduli.	It was the best of times. It was the worst of times. It was the age of wisdom. It was the age of foolishness. It was the epoch of belief. It was the epoch of incredulity.
1	A mde student rsts to olece after witnessing greed and orption wea hac on his life.	A model student resorts to violence after witnessing greed and corruption wreak havoc on his life.
2	A man wt blar isoer moves oe with his parents and makes a connection wh a spird widow, hih hes both of em heal in unique ways.	A man with bipolar disorder moves home with his parents and makes a connection with a spirited widow, which helps both of them heal in unique ways.
3	out African comedian Loyiso ol serves up filterfree uor as he ifs aut re, idnti, oitis, and a school prank gone embarngly wrong!	South African comedian Loyiso Gola serves up filterfree humor as he riffs about race, identity, politics, and a school prank gone embarrassingly wrong!

Figure 5.16 – Netflix random delete augmenting

Pluto does the same for the Twitter NLP data, as follows:

```
# use random delete augmentation technique
pluto.print_aug_char_random(pluto.df_twitter_data,
    action='delete', col_dest='clean_tweet',
    aug_name='Random Delete Augment')
```

The output is as follows:

	Random Delete Augment	Original
0	It was the es of mes. It was the rst of ime. It was the age of wsdo. It was the age of lishnss. It was the eoh of blif. It was the poc of incrdli.	It was the best of times. It was the worst of times. It was the age of wisdom. It was the age of foolishness. It was the epoch of belief. It was the epoch of incredulity.
1	el said	well said
2	@ ur went back to the sre, on to in out th one of the young empoys had resigned, due to srespetfl tretmt	@user went back to the store, only to find out that one of the young employees had resigned, due to disrespectful treatment
3	i am alue. # iam # osive # ffirion	i am valued. #iam #positive #affirmation

Figure 5.17 – Twitter random delete augmenting

Pluto uses the `print_aug_random()` wrapper function with `action` set to `substitute` for the **Netflix** NLP data, as follows:

```
# use random substitute augmentation technique
pluto.print_aug_char_random(pluto.df_netflix_data,
    action='substitute',
    col_dest='description',
    aug_name='Random Substitute Augment')
```

The output is as follows:

	Random Substitute Augment	Original
0	It was the b)sm of timas. It was the wo4st of tigks. It was the age of wi1dBm. It was the age of fo8l2shn9Fs. It was the epvvh of b2lies. It was the 4po5h of inc3e8uFiDy.	It was the best of times. It was the worst of times. It was the age of wisdom. It was the age of foolishness. It was the epoch of belief. It was the epoch of incredulity.
1	In his seRoHd standup lp9c#al, Daliem $oYa reminisces about his childhood, ponders MexM@Nn t+aditi%nY and points out a m6Vor problem with aocB.	In his second standup special, Daniel Sosa reminisces about his childhood, ponders Mexican traditions and points out a major problem with Coco.
2	A nameless widow sJggleC multiple jobs to sSppsAt her dvoghtWrs)eRisi$n to study ebroam, InMy to wssc0ver that she has ot9eg plans.	A nameless widow juggles multiple jobs to support her daughter's decision to study abroad, only to discover that she has other plans.
3	W_Dn a lonely tee^aOej ski0p his (ognJng c&aisms to sit in a lovely gaqdeP, he mem8s a mysterious oJter woman who shares his fHe@Mngs of alienation.	When a lonely teenager skips his morning classes to sit in a lovely garden, he meets a mysterious older woman who shares his feelings of alienation.

Figure 5.18 – Netflix random substitute augmenting

Pluto does the same for the **Twitter** NLP data, as follows:

```
# use random substitute augmentation technique
pluto.print_aug_char_random(pluto.df_twitter_data,
    action='substitute',
    col_dest='clean_tweet',
    aug_name='Random Substitute Augment')
```

The output is as follows:

	Random Substitute Augment	Original
0	It was the CeJt of tiXJs. It was the wVQst of tiFeX. It was the age of wXsdWm. It was the age of foOlisPn$ls. It was the eOoWh of belje$. It was the wpo!h of inFreduW9(y.	It was the best of times. It was the worst of times. It was the age of wisdom. It was the age of foolishness. It was the epoch of belief. It was the epoch of incredulity.
1	# op&ls baBkildq to call emergency opec meeting on proyucMiKn cut # bQo6 # &ilwer # gYlf # forex	#opecs barkindo to call emergency opec meeting on production cut #blog #silver #gold #forex
2	laEt n6ghK is the b0sq hahahahah # wAdleBvaywiedom # enough # date # hoAieF	last night is the best hahahahah #wednesdaywisdom #enough #date #homies
3	you r vdking us hlestkon why we vpteB 4u @ u^e3 8 yrs ago, but 0o^ey vpan7es ppl. u r$ld out. # lruDh	you r making us question why we voted 4u @user 8 yrs ago, but money changes ppl. u sold out. #truth

Figure 5.19 – Twitter random substitute augmenting

Pluto uses the `print_aug_random()` wrapper function with `action` set to `swap` for the **Netflix NLP** data, as follows:

```
# use random swap augmentation technique
pluto.print_aug_char_random(pluto.df_netflix_data,
    action='swap',
    col_dest='description',
    aug_name='Random Swap Augment')
```

The output is as follows:

	Random Swap Augment	Original
0	It was the ebts of tmise. It was the wsort of teims. It was the age of iwsdmo. It was the age of ofolishenss. It was the eophc of beilfe. It was the peohc of icnredluity.	It was the best of times. It was the worst of times. It was the age of wisdom. It was the age of foolishness. It was the epoch of belief. It was the epoch of incredulity.
1	Between scesne from his concert in So Apulso Theatro Munciipal, rapper and activist Emicida elcebratse the rhic leacgy of Black Brazilian ucultre.	Between scenes from his concert in Sao Paulo's Theatro Municipal, rapper and activist Emicida celebrates the rich legacy of Black Brazilian culture.
2	When an nuphapily amrrdie woman idsocvesr a man from her past has a role in a olcla theater production, shell do anything to reconnect twih him.	When an unhappily married woman discovers a man from her past has a role in a local theater production, she'll do anything to reconnect with him.
3	A computer anaytls ebcmeos a atrgte atfre she stumbles onto a ocnpsirayc via a ymsetriuos floppy disk, focrngi her to go on the run to celra her name.	A computer analyst becomes a target after she stumbles onto a conspiracy via a mysterious floppy disk, forcing her to go on the run to clear her name.

Figure 5.20 – Netflix random swap augmenting

Pluto does the same for the **Twitter** NLP data, as follows:

```
# use random swap augmentation technique
pluto.print_aug_char_random(pluto.df_twitter_data,
    action='swap',
    col_dest='clean_tweet',
    aug_name='Random Swap Augment')
```

The output is as follows:

	Random Swap Augment	Original
0	It was the best of itmse. It was the wsort of tmise. It was the age of wsidmo. It was the age of ofolisnhsse. It was the epcho of ebleif. It was the epoch of nicrdueltiy.	It was the best of times. It was the worst of times. It was the age of wisdom. It was the age of foolishness. It was the epoch of belief. It was the epoch of incredulity.
1	cnaont taste or smell anythign # ehplme # inomutstetohis # perfecttiming # weekend	cannot taste or smell anything #helpme #imnotusetothis #perfecttiming #weekend
2	child attacked by lailgaotr at walt disney woldr # lwatdisnyeowrdl	child attacked by alligator at walt disney world #waltdisneyworld
3	@ seur when its alreday utesyda and your wcw ndot even curr	@user when it's already tuesday and your wcw dont even curr

Figure 5.21 – Twitter random swap augmenting

> **Fun challenge**
>
> Here is a thought experiment: if the input text contains misspelled words and bad grammar, such as tweets, could correcting the spelling and grammar be a valid augmentation method?

Pluto has covered the **OCR, Keyboard**, and four modes of **Random** character augmentation techniques. The next step is augmenting words.

Word augmenting

At this point in the book, Pluto might think text augmentation is effortless, and it is true. We built a solid foundation layer in *Chapter 1* with an object-oriented class and learned how to extend the object as we learned about new augmentation techniques. In *Chapter 2*, Pluto added the functions for downloading any *Kaggle* real-world dataset, and *Chapters 3* and *4* gave us the wrapper function pattern. Therefore, at this point, Pluto reuses the methods and patterns to make the Python code concise and easy to understand.

The word augmentation process is similar to character augmentation. Pluto uses the same **Nlpaug** library to write the wrapper functions, which will invoke the `_print_aug_batch()` helper method. In particular, Pluto will cover the **Misspell, Split, Random, Synonyms, Antonyms**, and **Reserved** word augmenting techniques.

Let's start with the misspell augmentation technique.

Misspell augmenting

The misspell augmentation function uses a predefined dictionary to simulate spelling mistakes. The augmentation variable defines this as follows:

```
# define augmentation function variable definition
aug_func = nlpaug.augmenter.word.SpellingAug()
```

Pluto uses the `print_aug_word_misspell()` wrapper function on the **Netflix** NLP data, as follows:

```
# use word missplell augmentation
pluto.print_aug_word_misspell(pluto.df_netflix_data,
    col_dest='description',
    aug_name='Word Spelling Augment')
```

The output is as follows:

	Word Spelling Augment	Original
0	It was withe best of times. It wos Then worsest of times. Hit wa there's age of wisdom. It was the ago of foolishness. It was the epoch of believe. It was the epoch od incredulity.	It was the best of times. It was the worst of times. It was the age of wisdom. It was the age of foolishness. It was the epoch of belief. It was the epoch of incredulity.
1	Mr.. Peabody and Sherman host a zany latenight commedy showe Fron a swanky penthouse, winth timetraveling storical figures y a live audiance.	Mr. Peabody and Sherman host a zany late-night comedy show from a swanky penthouse, with timetraveling historical figures and a live audience.
2	As a wonem scours Hyderabad for her missing husband, she becames entangled yn a conspiracy the suggests theres moer top thi mysterie than meets trhe eye.	As a woman scours Hyderabad for her missing husband, she becomes entangled in a conspiracy that suggests there's more to the mystery than meets the eye.
3	In this biographical dramact, Britains King George VI stuggles wkth an embarrassed stutter utill he seeks hellp froom spech therapist Lionel Logue.	In this biographical drama, Britain's King George VI struggles with an embarrassing stutter until he seeks help from speech therapist Lionel Logue.

Figure 5.22 – Netflix misspell word augmenting

Pluto does the same for the **Twitter** NLP data, as follows:

```
# use word missplell augmentation
pluto.print_aug_word_misspell(pluto.df_twitter_data,
    col_dest='clean_tweet',
    aug_name='Word Spelling Augment')
```

The output is as follows:

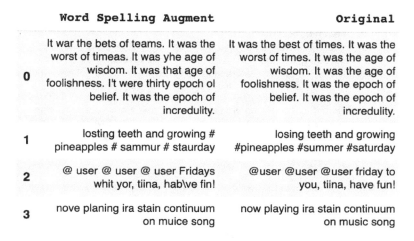

	Word Spelling Augment	Original
0	It war the bets of teams. It was the worst of timeas. It was yhe age of wisdom. It was that age of foolishness. I't were thirty epoch ol belief. It was the epoch of incredulity.	It was the best of times. It was the worst of times. It was the age of wisdom. It was the age of foolishness. It was the epoch of belief. It was the epoch of incredulity.
1	losting teeth and growing # pineapples # sammur # staurday	losing teeth and growing #pineapples #summer #saturday
2	@ user @ user @ user Fridays whit yor, tiina, hab\ve fin!	@user @user @user friday to you, tiina, have fun!
3	nove planing ira stain continuum on muice song	now playing ira stain continuum on music song

Figure 5.23 – Twitter misspell word augmenting

Similar to **Misspell** is the **Split** word augmentation technique.

Split augmenting

The split augmentation function randomly splits words into two tokens. The augmentation variable defines this as follows:

```
# define augmentation function variable definition
aug_func = nlpaug.augmenter.word.SplitAug()
```

Pluto uses the `print_aug_word_split()` wrapper function on the **Netflix** NLP data, as follows:

```
# use word split augmentation
pluto.print_aug_word_split(pluto.df_netflix_data,
    col_dest='description',
    aug_name='Word Split Augment')
```

The output is as follows:

	Word Split Augment	Original
0	It was the b est of tim es. It was the wo rst of tim es. It was the age of wi sdom. It was the age of fool ishness. It was the ep och of bel ief. It was the e poch of incre dulity.	It was the best of times. It was the worst of times. It was the age of wisdom. It was the age of foolishness. It was the epoch of belief. It was the epoch of incredulity.
1	To assert her independence from her rig id fa thers expec tations, a headst rong woman h ires a de sperate Los A ngeles man to pose as her f ianc in A frica.	To assert her independence from her rigid father's expectations, a headstrong woman hires a desperate Los Angeles man to pose as her fiancé in Africa.
2	A U. S. marsh als trou bling visions compro mise his investigation i nto the disappearance of a pat ient fr om a hospital for the cri minally i nsane.	A U.S. marshal's troubling visions compromise his investigation into the disappearance of a patient from a hospital for the criminally insane.
3	Clashes bo th on the st reet and in the bedro om p its the female he ad of the Orga nized Cr ime Unit aga inst the cocky le ader of a potsmuggling ring.	Clashes both on the street and in the bedroom pits the female head of the Organized Crime Unit against the cocky leader of a potsmuggling ring.

Figure 5.24 – Netflix split word augmenting

Pluto does the same for the **Twitter** NLP data, as follows:

```
# use word split augmentation
pluto.print_aug_word_split(pluto.df_twitter_data,
    col_dest='clean_tweet',
    aug_name='Word Split Augment')
```

The output is as follows:

	Word Split Augment	Original
0	It was the b est of ti mes. It was the w orst of ti mes. It was the age of w isdom. It was the age of fooli shness. It was the ep och of beli ef. It was the e poch of increduli ty.	It was the best of times. It was the worst of times. It was the age of wisdom. It was the age of foolishness. It was the epoch of belief. It was the epoch of incredulity.
1	# mirrorofthe witch ep9 ti me! !!	#mirrorofthewitch ep9 time!!!
2	# m agical rooster s imulation i w ant to climb the v ast expanse of mountains. it rea ched the lea kage wa	#magical rooster simulation i want to climb the vast expanse of mountains. it reached the leakage wa
3	why # cu stomers get ev en wh en youre po lite amp how to a void it @ user	why #customers get even when you're polite amp how to avoid it @user

Figure 5.25 – Twitter split word augmenting

After the split word method, Pluto presents the random word augmenting method.

Random augmenting

The random word augmentation method applies random behavior to the text with four parameters: **Swap**, **Crop**, **Substitute**, or **Delete**. The augmentation variable defines this as follows:

```
# define augmentation function variable definition
aug_func = nlpaug.augmenter.word.RandomWordAug(action=action)
```

Pluto uses the `print_aug_word_random()` wrapper function for swapping mode on the **Netflix** NLP data, as follows:

```
# use word random swap augmentation
pluto.print_aug_word_random(pluto.df_netflix_data,
    action='swap',
    col_dest='description',
    aug_name='Word Random Swap Augment')
```

The output is as follows:

	Word Random Swap Augment	Original
0	It was the times best of. was It the worst of times. It the age was of wisdom It. the was age of foolishness. was It the epoch of belief. It was the of epoch. incredulity	It was the best of times. It was the worst of times. It was the age of wisdom. It was the age of foolishness. It was the epoch of belief. It was the epoch of incredulity.
1	Herself talented a, OEi artist works her with father, Tetsuzo, later known as Hokusai, the on woodblock prints that make would Edo worldwide famous.	Herself a talented artist, OEi works with her father, Tetsuzo, later known as Hokusai, on the woodblock prints that would make Edo famous worldwide.
2	Massage a therapist gets in over he his head when partners with a charismatic pal childhood the in lucrative but shady of business global arms dealing.	A massage therapist gets in over his head when he partners with a charismatic childhood pal in the lucrative but shady business of global arms dealing.
3	Nave a social worker a brings 10yearold child into her home to the rescue from girl abusive parents only to she learn that what isnt she seems.	A naive social worker brings a 10-year-old child into her home to rescue the girl from abusive parents only to learn that she isn't what she seems.

Figure 5.26 – Netflix random swap word augmenting

Pluto does the same for the **Twitter** NLP data, as follows:

```
# use word random swap augmentation
pluto.print_aug_word_random(pluto.df_twitter_data,
    action='swap',
    col_dest='clean_tweet',
    aug_name='Word Random Swap Augment')
```

The output is as follows:

	Word Random Swap Augment	Original
0	It was best the of times. was It the worst of times. It the was age of wisdom. It was the age of foolishness. It was the epoch belief of. was It the epoch incredulity of.	It was the best of times. It was the worst of times. It was the age of wisdom. It was the age of foolishness. It was the epoch of belief. It was the epoch of incredulity.
1	@ user these wever are ppl love who the rep need pay open to effing newpaper a. . and. im a republican! waiting4trump #	@user wever these ppl are who love the rep pay need to open a effing newpaper...and im a republican! #waiting4trump
2	@ how user a to become # successful ex amp #. singlemom # # aga3 dati	@user how to become a #successful amp ex#singlemom. #aga3 #dati
3	Service bubble! bubbles # # # brunch cheers # # sinema # fathersday nashville # happiness	bubble service! #bubbles #cheers #brunch #sinema #fathersday #nashville #happiness

Figure 5.27 – Twitter random swap word augmenting

Pluto uses the `print_aug_word_random()` wrapper function for cropping mode on the **Netflix** NLP data, as follows:

```
# use word random crop augmentation
pluto.print_aug_word_random(pluto.df_netflix_data,
    action='crop',
    col_dest='description',
    aug_name='Word Random Crop Augment')
```

The output is as follows:

	Word Random Crop Augment	Original
0	It was the best of times. It was the worst of times. It was the age of wisdom. It of belief. It was the epoch of incredulity.	It was the best of times. It was the worst of times. It was the age of wisdom. It was the age of foolishness. It was the epoch of belief. It was the epoch of incredulity.
1	After a life of crime, a notorious kidnapper tries to to win the heart of the woman he loves.	After a life of crime, a notorious kidnapper tries to change his ways and turn over a new leaf to win the heart of the woman he loves.
2	Richard accepts a bet that he cant find a companion app hes using includes a 45day contract.	Richard accepts a bet that he can't find a companion to a friend's wedding, not knowing the dating app hes using includes a 45-day contract.
3	After Marcos reinvent himself as a modern man with the help of a childhood friend and an online guru.	After Marcos is dumped by his girlfriend, he attempts to reinvent himself as a modern man with the help of a childhood friend and an online guru.

Figure 5.28 – Netflix random crop word augmenting

Pluto does the same for the **Twitter** NLP data, as follows:

```
# use word random swap augmentation
pluto.print_aug_word_random(pluto.df_twitter_data,
    action='crop',
    col_dest='clean_tweet',
    aug_name='Word Random Crop Augment')
```

The output is as follows:

	Word Random Crop Augment	Original
0	It was the best of times. It was the worst of times. It was the age of wisdom the epoch of belief. It was the epoch of incredulity.	It was the best of times. It was the worst of times. It was the age of wisdom. It was the age of foolishness. It was the epoch of belief. It was the epoch of incredulity.
1	user hit on foot so until the cars around us chill we could all end	@user he was hit on foot so until the cars around us chill, we could all end up like him
2	how am i going kid when i cant even play lax myself	how am i going to give lacrosse lessons to this kid when i cant even play lax myself
3	a successful day of hunting # johnxsafaris # eastcapehunting # family	after a successful day of hunting #johnxsafaris #eastcapehunting #family #proud #africa

Figure 5.29 – Twitter random crop word augmenting

So, Pluto has described the **Swap** and **Crop** word augmentation methods but not the **Substitute** and **Delete** ones. This is because they are similar to the character augmenting functions and are in the Python Notebook. Next on the block is synonym augmenting.

Synonym augmenting

The synonym augmentation function substitutes words with synonyms from a predefined database. **WordNet** and **PPBD** are two optional databases. The augmentation variable defines this process as follows:

```
# define augmentation function variable definition
aug_func = nlpaug.augmenter.word.SynonymAug(
    aug_src='wordnet')
```

Pluto uses the `print_aug_word_synonym()` wrapper function on the **Netflix** NLP data, as follows:

```
# use word synonym augmentation
pluto.print_aug_word_synonym(pluto.df_netflix_data,
    col_dest='description',
    aug_name='Synonym WordNet Augment')
```

The output is as follows:

	Synonym WordNet Augment	Original
0	Information technology be the best of times. It was the worst of metre. It was the years of wisdom. It be the geezerhood of foolishness. It was the epoch of belief. It was the date of reference of incredulity.	It was the best of times. It was the worst of times. It was the age of wisdom. It was the age of foolishness. It was the epoch of belief. It was the epoch of incredulity.
1	Following a disaster, a female parent and her teen son move to a relatives vacant holiday home, where eerie and unexplained force out conspire against them.	Following a tragedy, a mother and her teen son move to a relatives vacant vacation home, where eerie and unexplained forces conspire against them.
2	After her logos dies in an stroke, a sound consultant is force to prove her innocence when shes criminate of his murder.	After her son dies in an accident, a legal consultant is forced to prove her innocence when she's accused of his murder.
3	After 16yearold Cyntoia Brown make up sentenced to life in prison house, interrogative sentence astir her past, physiology and the law itself yell her guilt into question.	After 16-year-old Cyntoia Brown is sentenced to life in prison, questions about her past, physiology and the law itself call her guilt into question.

Figure 5.30 – Netflix synonym word augmenting

It is interesting and funny that the synonym of *It* is **Information Technology** for the control text. Mr. Dickens, who wrote Tale of Two Cities in 1859, could never have known that IT is a popular acronym for information technology. Pluto does the same for the **Twitter** NLP data, as follows:

```
# use word synonym augmentation
pluto.print_aug_word_synonym(pluto.df_twitter_data,
    col_dest='clean_tweet',
    aug_name='Synonym WordNet Augment')
```

The output is as follows:

	Synonym WordNet Augment	Original
0	It was the best of time. It was the bad of times. It constitute the age of wisdom. It be the long time of foolishness. It was the epoch of feeling. Information technology was the epoch of disbelief.	It was the best of times. It was the worst of times. It was the age of wisdom. It was the age of foolishness. It was the epoch of belief. It was the epoch of incredulity.
1	im sure the @ user testament withal suppo him amp the right field to have a gunman. # wakeupamerica!!	i'm sure the @user will still suppo him amp the right to have a gun. #wakeupamerica !!
2	# coffe # cake # tea # teapay fourth dimension @ reggio coffee berry house	#coffee #cake #tea #teapay time @ reggio coffee house
3	hardcore vinyls the best asian site	hardcore vinyls the best asian site

Figure 5.31 – Twitter synonym word augmenting

When there are synonyms, you will also find antonyms.

Antonym augmenting

The antonym augmentation function randomly replaces words with antonyms. The augmentation variable defines this as follows:

```
# define augmentation function variable definition
aug_func = nlpaug.augmenter.word.AntonymAug()
```

Pluto uses the print_aug_word_antonym() wrapper function on the **Netflix** NLP data, as follows:

```
# use word antonym augmentation
pluto.print_aug_word_antonym(pluto.df_netflix_data,
    col_dest='description',
    aug_name='Antonym Augment')
```

The output is as follows:

	Antonym Augment	Original
0	It differ the evil of times. It differ the best of times. It differ the age of wisdom. It differ the age of foolishness. It differ the epoch of belief. It differ the epoch of incredulity.	It was the best of times. It was the worst of times. It was the age of wisdom. It was the age of foolishness. It was the epoch of belief. It was the epoch of incredulity.
1	Comedy duo Thomas Middleditch and Ben Schwartz unbend large ideas into epically funny stories in this series of partly improvised comedy specials.	Comedy duo Thomas Middleditch and Ben Schwartz turn small ideas into epically funny stories in this series of completely improvised comedy specials.
2	The cast and crew of the 1981 Broadway unmusical Merrily We Roll Along recall joy and heartbreak during the production of a surefire hit that wasnt.	The cast and crew of the 1981 Broadway musical Merrily We Roll Along recall joy and heartbreak during the production of a surefire hit that wasn't.
3	Extreme pressure from his father to excel at school during childhood lack safe psychological effects on a brilliant but suicidal mans adult life.	Extreme pressure from his father to excel at school during childhood has dangerous psychological effects on a brilliant but suicidal man's adult life.

Figure 5.32 – Netflix antonym word augmenting

Pluto does the same for the **Twitter** NLP data, as follows:

```
# use word antonym augmentation
pluto.print_aug_word_antonym(pluto.df_twitter_data,
    col_dest='clean_tweet',
    aug_name='Antonym Augment')
```

The output is as follows:

	Antonym Augment	Original
0	It differ the worst of times. It differ the unregretful of times. It differ the age of wisdom. It differ the age of foolishness. It differ the epoch of belief. It differ the epoch of incredulity.	It was the best of times. It was the worst of times. It was the age of wisdom. It was the age of foolishness. It was the epoch of belief. It was the epoch of incredulity.
1	# visiting gorilla simulator you obviate to unmake to adapt to the environment. the need to tear the city. materi	#visiting gorilla simulator you need to do to adapt to the environment. the need to tear the city. materi
2	@ user whoopppp stop take away me all idle up # ska # reggae	@user whoopppp stop getting me all worked up #ska #reggae
3	i differ stunning. # iam # negative # affirmation	i am stunning. #iam #positive #affirmation

Figure 5.33 – Twitter antonym word augmenting

After synonyms and antonyms, which are automated, reserved word augmentation requires a manual word list.

Reserved word augmenting

The reserved word augmentation method swaps target words where you define a word list. It is the same as synonyms, except the terms are created manually. Pluto uses the **Netflix** and **Twitter** word cloud diagrams, *Figures 5.8* and *5.9*, to select the top three reoccurring words in the NLP datasets. The augmentation variable defines this process as follows:

```
# define augmentation function
aug_func = nlpaug.augmenter.word.ReservedAug(
    reserved_tokens=reserved_tokens)
# define control sentence reserved words
pluto.reserved_control = [['wisdom', 'sagacity',
    'intelligence', 'prudence'],
    ['foolishness', 'folly', 'idiocy', 'stupidity']]
# define Netflix reserved words
pluto.reserved_netflix = [['family','household', 'brood',
    'unit', 'families'],
    ['life','existance', 'entity', 'creation'],
    ['love', 'warmth', 'endearment','tenderness']]
pluto.reserved_netflix = pluto.reserved_control +
    pluto.reserved_netflix
# define Twitter reserved words
```

```
pluto.reserved_twitter = [['user', 'users', 'customer',
    'client','people','member','shopper'],
    ['happy', 'cheerful', 'joyful', 'carefree'],
    ['time','clock','hour']]
pluto.reserved_twitter = pluto.reserved_control +
    pluto.reserved_twitter
```

Pluto uses the `print_aug_word_reserved()` wrapper function on the **Netflix** NLP data, as follows:

```
# use word reserved augmentation
pluto.print_aug_word_reserved(pluto.df_netflix_data,
    col_dest='description',
    reserved_tokens=pluto.reserved_netflix)
```

The output is as follows:

	Netflix Reserved word augment	Original
0	It was the best of times. It was the worst of times. It was the age of intelligence. It was the age of idiocy. It was the epoch of belief. It was the epoch of incredulity.	It was the best of times. It was the worst of times. It was the age of wisdom. It was the age of foolishness. It was the epoch of belief. It was the epoch of incredulity.
1	The elite real estate brokers at The Oppenheim Group sell the luxe existance to affluent buyers in LA. The drama ramps up when a new agent joins the team.	The elite real estate brokers at The Oppenheim Group sell the luxe life to affluent buyers in LA. The drama ramps up when a new agent joins the team.
2	In this dramatization, the Virgin Mary works a miracle on a girl in 1623 Mexico. Four centuries later, a brood make a pilgrimage for their own child.	In this dramatization, the Virgin Mary works a miracle on a girl in 1623 Mexico. Four centuries later, a family make a pilgrimage for their own child.
3	Evidence found on the body of a homicide victim sparks hope in a prosecutor that his sister who disappeared 25 years earlier could still be alive.	Evidence found on the body of a homicide victim sparks hope in a prosecutor that his sister who disappeared 25 years earlier could still be alive.

Figure 5.34 – Netflix reserved word augmenting

Notice the words **wisdom** and **foolishness** are substituted with **Intelligence** and **idiocy, life** with **existance**, and **family** with **brood**. Pluto does the same for the **Twitter** NLP data, as follows:

```
# use word reserved augmentation
pluto.print_aug_word_reserved(pluto.df_twitter_data,
    col_dest='clean_tweet',
    reserved_tokens=pluto.reserved_twitter)
```

The output is as follows:

	Twitter Reserved word augment	Original
0	It was the best of times. It was the worst of times. It was the age of sagacity. It was the age of idiocy. It was the epoch of belief. It was the epoch of incredulity.	It was the best of times. It was the worst of times. It was the age of wisdom. It was the age of foolishness. It was the epoch of belief. It was the epoch of incredulity.
1	@ people i am thankful for freedom of speech. # thankful # positive	@user i am thankful for freedom of speech. #thankful #positive
2	there is a mona lisa style to this mans face # sad @ customer	there is a mona lisa style to this man's face #sad @user
3	quality hour with my man. # homedate # weekend # loved	quality time with my man. #homedate #weekend #loved

Figure 5.35 – Twitter reserved word augmenting

Notice the words **wisdom** and **foolishness** are substituted with **sagacity** and **idiocy**, and **user** with **people** and **customer**.

Reserved Word augmenting is the last word augmentation method of this chapter. Pluto has covered **Misspell**, **Split**, **Random**, **Synonym**, **Antonym**, and **Reserved Word** augmentation, but these are only some of the possible word augmentation techniques you can use.

Fun challenge

The challenge is to use the Augly library or the NLTK, Gensim, or Textblob libraries to write a new wrapper function. It is relatively easy. The first step is to copy a wrapper function, such as the `print_aug_keyboard()` function. The second and last step is to replace `aug_func = nlpaug.augmenter.char.KeyboardAug()` with `aug_func = augly.text.functional.simulate_typos()`. There are more parameters in the Augly function. A hint is to use the `augly.text.functional.simulate_typos?` command to display the function documentation.

The **Nlpaug** library and other text augmentation libraries, such as **NLTK, Gensim, Textblob**, and **Augly,** have additional text augmentation methods. In addition, newly published scholarly papers are an excellent source in which to discover new text augmentation techniques.

Let's summarize this chapter.

Summary

At first glance, text augmentation seems counterintuitive and problematic because the techniques inject errors into the text. Still, DL based on CNNs or RNNs recognizes patterns regardless of a few misspellings or synonym replacements. Furthermore, many published scholarly papers have described the benefits of text augmentation to increase prediction or forecast accuracy.

In *Chapter 5*, you learned about three **Character** augmentation techniques, **OCR, Keyboard**, and **Random**. In addition, the six **Word** augmentation techniques are the **Misspell, Split, Random, Synonyms, Antonyms**, and **Reserved** words. There are more text augmentation methods in the Nlgaug, NLTK, Gensim, TextBlob, and Augly libraries.

Implementing the text augmentation methods using a Python Notebook is deceptively simple. This is because Pluto built a solid foundation layer in *Chapter 1* with an object-oriented class and learned how to extend the object with **decorator** as he discovered new augmentation techniques. In *Chapter 2*, Pluto added the functions for downloading any *Kaggle* real-world dataset, and *Chapters 3* and *4* gave us the wrapper function pattern. Therefore, in this chapter, Pluto reused the methods and patterns to make the Python Notebook code concise and easy to understand.

Throughout the chapter, there are *Fun facts* and *Fun challenges*. Pluto hopes you will take advantage of them and expand your experience beyond the scope of this chapter.

The next chapter will delve deeper into text augmentation using machine learning methods. Ironically, the goal of text augmentation is to make machine learning and DL predict and forecast accurately, and we will use the same AI system to increase the efficiency of text augmentation. It is a circular logic or cyclical process.

Pluto is waiting for you in the next chapter, *Text Augmentation with Machine Learning*.

6
Text Augmentation with Machine Learning

Text augmentation with **machine learning (ML)** is an advanced technique compared to the standard text augmenting methods we covered in the previous chapter. Ironically, text augmentation aims to improve ML model accuracy, but we used a pre-trained ML model to create additional training NLP data. It's a circular process. ML coding is not in this book's scope, but understanding the difference between using libraries and ML for text augmentation is beneficial.

Augmentation libraries, whether for image, text, or audio, follow the traditional programming methodologies with structure data, loops, and conditional statements in the algorithm. For example, as shown in *Chapter 5*, the pseudocode for implementing the _print_aug_reserved() method could be as follows:

```
# define synonym words, pseudo-code
reserved = [['happy', 'joyful', 'cheerful'],
  ['sad', 'sorrowful', 'regretful']]
# substitute the word with its synonym, pseudo-code
for i, word in (input_text)
  for set_word in (reserved)
    for i, syn in set_word
      if (syn == word)
        input_text[i] = set_word[i+1]
```

The happy path code does not cover error checking, but the salient point is that the library's function follows the standard sequential coding method.

On the other hand, ML is based on one of the 13 known ML algorithms, including **deep learning (DL)** (or **artificial neural networks**), **Bidirectional Encoder Representations from Transformers (BERT)**, **linear regression**, **random forest**, **naive Bayes**, and **gradient boosting**. The key to ML is that the system *learns* and not programs. DL uses the Universal Approximation theory, gradient descent, transfer learning, and hundreds of other techniques. ANNs have millions to billions of neural nodes – for example, OpenAI GPT3 has 96 layers and 175 billion nodes. The central point is that ML has no familiarity with the `_print_aug_reserved()` pseudocode algorithm.

The following is a representation of a DL architecture for image classification. It illustrates the difference between a procedural approach and the Neural Network algorithm. This figure was created from **Latex** and the **Overleaf** cloud system. The output is as follows:

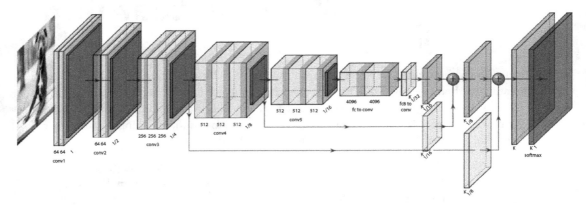

Figure 6.1 – Representation of a DL model

The Overleaf project and its code are from Mr. Duc Haba's public repository, and the URL is https://www.overleaf.com/project/6369a1eaba583e7cd423171b. You can clone and hack the code to display other AI models.

This chapter will cover text augmentation with ML, and in particular, the following topics:

- Machine learning models
- Word augmenting
- Sentence augmenting
- Real-world NLP datasets
- Reinforcing your learning through the Python Notebook

Let's briefly describe the ML models used in the Python wrapper function code.

Machine learning models

In this chapter, the text augmentation wrapper functions use ML to generate new text for training the ML model. Understanding how these models are built is not in scope, but a brief description of these ML models and their algorithms is necessary. The Python wrapper functions will use the following ML models under the hood:

- Tomáš Mikolov published the NLP algorithm using a neural network named **Word2Vec** in 2013. The model can propose synonym words from the input text.

- The **Global Vectors for Word Representation (GloVe)** algorithm was created by Jeffrey Pennington, Richard Socher, and Christopher D. Manning in 2014. It is an unsupervised learning NLP algorithm for representing words in vector format. The results are a linear algorithm that groups the closest neighboring words.

- **Wiki-news-300d-1M** is a pre-trained ML model that uses the **fastText** open source library. It was trained on 1 million words from Wikipedia 2017 articles, the UMBC WebBase corpus, which consists of over 3 billion words, and the Statmt.org news dataset, which consists of over 16 billion tokens. T. Mikolov, E. Grave, P. Bojanowski, C. Puhrsch, and A. Joulin introduced **Wiki-news-300d-1M** in their *Advances in Pre-Training Distributed Word Representations* paper. The license is the Creative Commons Attribution-Share-Alike License 3.0.

- **GoogleNews-vectors-negative300** is a pre-trained **Word2Vec** model that uses the Google News dataset, which contains about 100 billion words and 300 dimensions.

- Google introduced the **transformer** neural network algorithm in 2017. Recent cutting-edge breakthroughs in NLP and computer vision are from the transformer model.

- The **BERT** model was introduced by Jacob Devlin, Ming-Wei Chang, Kenton Lee, and Kristina Toutanova in 2018. It is specialized in language inference and prediction.

- **RoBERTa** is an optimized algorithm for the self-supervised NLP model. It is a model built on top of BERT. It excels in performance on many NLP inferences. Meta AI published RoBERTa in 2019.

- Facebook's **wmt19-en-de** and **wmt19-de-en** are pre-trained NLP models from *HuggingFace* for translating from English to German (Deutsch) and back. It was made publicly available in 2021.

- Facebook's **wmt19-en-ru** and **wmt19-ru-en** are pre-trained NLP models from *HuggingFace* for translating from English to Russian (Русский) and back. It was made publicly available in 2021.

- **XLNet** is a transformer-XL pre-trained model that was made publicly available by Zhilin Yang, Zihang Dai, Yiming Yang, Jaime Carbonell, Russ R. Salakhutdinov, and Quoc V. Le on *HuggingFace* in 2021. It was published in the scholarly paper *XLNet: Generalized Autoregressive Pretraining for Language Understanding*.

- The **Generative Pre-trained Transformer 2 (GPT-2)** algorithm is an open source AI that was published by OpenAI in 2019. The model excels in writing feedback questions and answers and generating text summarization of an article. It is at the level of actual human writing.

- The **T5** and **T5X** models use the text-to-text transformer algorithm. They were trained on a massive corpus. Colin Raffel, Noam Shazeer, Adam Roberts, Katherine Lee, Sharan Narang, Michael Matena, Yanqi Zhou, Wei Li, and Peter J. Liu introduced T5 in their paper *Exploring the Limits of Transfer Learning with a Unified Text-to-Text Transformer* in 2020.

> **Fun fact**
>
> Generative AI, when using a transformer model, such as OpenAI's GPT-3, GPT-4, or Google Bard, can write as well or better than a human writer.

Now that we know about some of the ML models, let's see which augmenting function uses which ML models.

Word augmenting

In this chapter, the word augmenting techniques are similar to the methods from *Chapter 5*, which used the **Nlpaug** library. The difference is that rather than Python libraries, the wrapper functions use powerful ML models to achieve remarkable results. Sometimes, the output or rewritten text is akin to human writers.

In particular, you will learn four new techniques and two variants each. Let's start with Word2Vec:

- The **Word2Vec** method uses the neural network NLP Word2Vec algorithm and the GoogleNews-vectors-negative300 pre-trained model. Google trained it using a large corpus containing about 100 billion words and 300 dimensions. Substitute and insert are the two mode variants.

- The **BERT** method uses Google's transformer algorithm and BERT pre-trained model. Substitute and insert are the two mode variants.

- The **RoBERTa** method is a variation of the BERT model. Substitute and insert are the two mode variants.

- The last word augmenting technique that we'll look at in this chapter is **back translation** using Facebook's (Meta's) pre-trained translation model. It translates the input English text into a different language and back to English. The two variants we'll look at involve translating from English into German (Deutsch) and back to English using the **facebook/wmt19-en-de** and **facebook/wmt19-de-en** models, and from English to Russian (Русский) and back to English using the **facebook/wmt19-en-ru** and **facebook/wmt19-ru-en** models.

It will be easier to understand this by reading the output from the word wrapper functions, but before we do, let's describe sentence augmenting.

Sentence augmenting

Augmenting at the sentence level is a powerful concept. It was not possible 5 years ago. You had to be working in an ML research company or a billionaire before accessing these acclaimed pre-trained models. Some transformer and **large language models** (**LLMs**) became available in 2019 and 2020 as open source, but they are generally for research. Convenient access to online AI servers via a GPU was not widely available at that time. The LLM and pre-trained models have recently become publicly accessible for incorporating them into your projects, such as the HuggingFace website. The salient point is that for independent researchers or students, LLM and pre-trained models only became accessible in mid-2021.

The sentence and word augmenting methods that use ML can't be done dynamically as with methods using the **Nlpaug** library. In other words, you have to write and save the augmented text to your local or cloud disk space. The primary reason is that the augmentation step takes too long per training cycle. The upside is that you can increase the original text by 20 to 100 times its size.

In particular, we will cover the following techniques:

- Summarizing text using the **T5** NLP algorithm.
- **Sequence** and **Sometimes** are two sentence flow methods. The flow methods use a combination of the **GloVe** and **BERT** NLP algorithms.

For the sentence augmentation techniques, they are easier to understand by reading the output of the wrapper functions using real-world NLP datasets as input text. Thus, the following section is about writing wrapper functions with Python code to gain insight into sentence augmenting, but first, let's download the real-world NLP datasets.

Real-world NLP datasets

This chapter will use the same Netflix and Twitter real-world NLP datasets from *Chapter 5*. In addition, both datasets have been vetted, cleaned, and stored in the `pluto_data` directory in this book's GitHub repository. The startup sequence is similar to the previous chapters. It is as follows:

1. Clone the Python Notebook and Pluto.
2. Verify Pluto.
3. Locate the NLP data.
4. Load the data into pandas.
5. View the data.

Let's start with the Python Notebook and Pluto.

Python Notebook and Pluto

Start by loading the `data_augmentation_with_python_chapter_6.ipynb` file into Google Colab or your chosen Jupyter Notebook or JupyterLab environment. From this point onward, we will only display code snippets. The complete Python code can be found in the Python Notebook.

The next step is to clone the repository. We will reuse the code from *Chapter 5*. The `!git` and `%run` statements are used to instantiate Pluto:

```
# clone Packt GitHub repo.
!git clone 'https://github.com/PacktPublishing/Data-
Augmentation-with-Python'
# Instantiate Pluto
%run 'Data-Augmentation-with-Python/pluto/pluto_chapter_5.py'
```

The output is as follows:

```
---------------------------- : ----------------------------
                    Hello from class : <class '__main__.
PacktDataAug'> Class: PacktDataAug
                         Code name : Pluto
                         Author is : Duc Haba
---------------------------- : ----------------------------
```

The following setup step is checking if Pluto loaded correctly.

Verify

The following command asks Pluto to display his status:

```
# Am I alive?
pluto.say_sys_info()
```

The output will be as follows or similar, depending on your system:

```
---------------------------- : ----------------------------
              System time : 2022/11/09 05:31
                 Platform : linux
    Pluto Version (Chapter) : 5.0
            Python (3.7.10) : actual: 3.7.15 (default, Oct 12 2022,
19:14:55) [GCC 7.5.0]
            PyTorch (1.11.0) : actual: 1.12.1+cu113
            Pandas (1.3.5) : actual: 1.3.5
               PIL (9.0.0) : actual: 7.1.2
        Matplotlib (3.2.2) : actual: 3.2.2
                CPU count : 12
                CPU speed : NOT available
---------------------------- : ----------------------------
```

Pluto showed that he is from *Chapter 5* (version 5.0), which is correct. In addition, the cleaned NLP Twitter and Netflix datasets are in the ~/Data-Augmentation-with-Python/pluto_ data directory.

Real-world NLP data

Pluto is using the clean versions of the data without profanity from *Chapter 5*. They are the Netflix and Twitter NLP datasets from the Kaggle website. The clean datasets were saved in this book's GitHub repository. Thus, Pluto does not need to download them again. Still, you can download them or other real-world datasets by using the fetch_kaggle_dataset() function. Pluto locates the cleaned NLP datasets with the following commands:

```
# check to see the files are there
f = 'Data-Augmentation-with-Python/pluto_data'
!ls -la {f}
```

The output is as follows:

```
drwxr-xr-x 2 root root    4096 Nov 13 06:07 .
drwxr-xr-x 6 root root    4096 Nov 13 06:07 ..
-rw-r--r-- 1 root root 3423079 Nov 13 06:07 netflix_data.csv
-rw-r--r-- 1 root root 6072376 Nov 13 06:07 twitter_data.csv
```

> **Fun fact**
>
> Pluto gets lazy, and instead of using a Python library and coding it in Python, he cheats by dropping down to the Linux Bash command-line code. The exclamation character (!) allows the Python Notebook to backdoor the kernel, such as via !ls -la on Linux or !dir on Windows. You can use any OS command-line code. Still, it is not portable code because the commands for Windows, iOS, Linux, Android, and other OSs that support web browsers such as Safari, Chrome, Edge, and Firefox are different.

The next step is to load the data into Pluto's buddy, pandas.

Pandas

Pluto reuses the fetch_df() method from *Chapter 2* to load the data into pandas. The following commands import the real-world Netflix data into pandas:

```
# import to Pandas
f = 'Data-Augmentation-with-Python/pluto_data/netflix_data.csv'
pluto.df_netflix_data = pluto.fetch_df(f,sep='~')
```

Similarly, the commands for loading the real-world Twitter data are as follows:

```
# import to Pandas
f = 'Data-Augmentation-with-Python/pluto_data/twitter_data.csv'
pluto.df_twitter_data = pluto.fetch_df(f,sep='~')
```

> **Fun challenge**
> Pluto challenges you to find and download two additional NLP data from the Kaggle website. Hint: use Pluto's `fetch_kaggle_dataset()` function. Import it into pandas using the `fetch_df()` function.

Now that Pluto has located and imported the data into pandas, the last step in loading the data sequence is to view and verify the data.

Viewing the text

The `draw_word_count()` and `draw_null_data()` methods help us understand the NLP data, and Pluto recommends revisiting *Chapter 5* to view those Netflix and Twitter graphs. A more colorful and fun method is to use the `draw_word_cloud()` function.

Pluto draws the Netflix word cloud infographic graph with the following command:

```
# draw infographic plot
pluto.draw_text_wordcloud(pluto.df_netflix_data.description,
  xignore_words=wordcloud.STOPWORDS,
  title='Word Cloud: Netflix Movie Review')
```

The output is as follows:

Figure 6.2– Netflix word cloud

Similarly, Pluto displays the Twitter word cloud using the following commands:

```
# draw infographic plot
pluto.draw_text_wordcloud(pluto.df_twitter_data.clean_tweet,
    xignore_words=wordcloud.STOPWORDS,
    title='Word Cloud: Twitter Tweets')
```

The output is as follows:

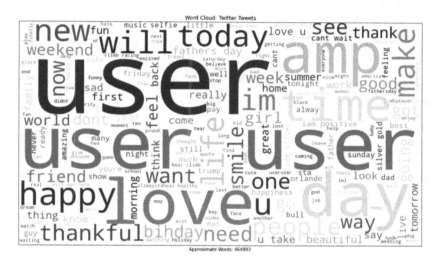

Figure 6.3 – Twitter word cloud

Along with the real-world NLP data, Pluto uses the first few lines of the *Tale of Two Cities*, by Charles Dickens, as the control text. In this chapter, Pluto will extend the control text to the first page of Mr. Dickens' book, the *Moby Dick* book, by Melville, and the *Alice in Wonderland* book, by Carroll. These books are in public domain, as defined in Project Gutenberg.

The varibles are `pluto.orig_text`, `pluto.orig_dickens_page`, `pluto.orig_melville_page`, and `pluto.orig_carroll_page`, respectively.

> **Fun fact**
>
> ML is good at altering text in typical human writing but modifying the masterworks is borderline criminal. Pluto seeks only to illustrate the augmentation concepts and never to bastardize the classics. It is in the name of science.

You have loaded the Python Notebook, instantiated Pluto, accessed the cleaned NLP real-world data, and verified it with the word cloud infographic. Now, it is time to write and hack Python code to gain a deeper insight into word and sentence augmentation with ML.

Reinforcing your learning through the Python Notebook

Even though NLP ML is highly complex, the implementation for the wrapper code is deceptively simple. This is because of Pluto's structured object-oriented approach. First, we created a base class for Pluto in *Chapter 1* and used the decorator to add a new method as we learned new augmentation concepts. In *Chapter 2*, Pluto learned to download any of the thousands of real-world datasets from the Kaggle website. *Chapters 3* and *4* introduced the wrapper functions process using powerful open source libraries under the hood. Finally, *Chapter 5* put forward the text augmentation concepts and methods when using the **Nlpaug** library.

Therefore, building upon our previous knowledge, the wrapper functions in this chapter use the powerful NLP ML pre-trained model to perform the augmentations.

In particular, this chapter will present wrapper functions and the augmenting results for the Netflix and Twitter real-world datasets using the following techniques:

- **Word2Vec** word augmenting
- **BERT** and **Transformer** word augmenting
- **RoBERTa** augmenting
- **Back translation**
- **T5** augmenting
- **Sequential** and **Sometime** augmenting

Let's start with Word2Vec.

Word2Vec word augmenting

The `print_aug_ai_word2vec()` wrapper function's key parameters are as follows:

```
# code snippet for print_aug_ai_word2vec()
model_type = 'word2vec',
model_path = 'GoogleNews-vectors-negative300.bin'
action = 'insert'    # or 'substitute'
nlpaug.augmenter.word.WordEmbsAug(model_type,
  model_path,
  action)
```

The full functions can be found in the Python Notebook. Pluto uses the real-world NLP Netflix data to test the function, as follows:

```
# augment using word2vec
pluto.print_aug_ai_word2vec(pluto.df_netflix_data,
  col_dest='description',
```

```
action='insert',
aug_name='Word2Vec-GoogleNews Word Embedding Augment')
```

> **Fun fact**
>
> When you run a wrapper function, new data is randomly selected and processed. Thus, it would be best if you run the wrapper function repeatedly to see different movie reviews from the Netflix dataset or tweets from the Twitter dataset.

The output is as follows:

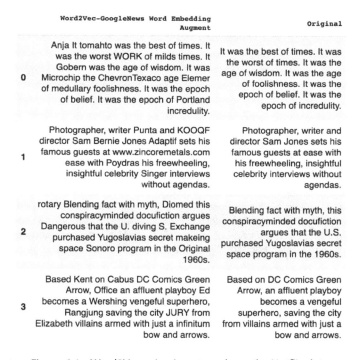

	Word2Vec-GoogleNews Word Embedding Augment	Original
0	Anja It tomahto was the best of times. It was the worst WORK of milds times. It Gobern was the age of wisdom. It was Microchip the ChevronTexaco age Elemer of medullary foolishness. It was the epoch of belief. It was the epoch of Portland incredulity.	It was the best of times. It was the worst of times. It was the age of wisdom. It was the age of foolishness. It was the epoch of belief. It was the epoch of incredulity.
1	Photographer, writer Punta and KOOQF director Sam Bernie Jones Adaptif sets his famous guests at www.zincoremetals.com ease with Poydras his freewheeling, insightful celebrity Singer interviews without agendas.	Photographer, writer and director Sam Jones sets his famous guests at ease with his freewheeling, insightful celebrity interviews without agendas.
2	rotary Blending fact with myth, Diomed this conspiracyminded docufiction argues Dangerous that the U. diving S. Exchange purchased Yugoslavias secret makeing space Sonoro program in the Original 1960s.	Blending fact with myth, this conspiracyminded docufiction argues that the U.S. purchased Yugoslavias secret space program in the 1960s.
3	Based Kent on Cabus DC Comics Green Arrow, Office an affluent playboy Ed becomes a Wershing vengeful superhero, Rangjung saving the city JURY from Elizabeth villains armed with just a infinitum bow and arrows.	Based on DC Comics Green Arrow, an affluent playboy becomes a vengeful superhero, saving the city from villains armed with just a bow and arrows.

Figure 6.4 – Word2Vec using insert mode on the Netflix data

In *Figure 6.3*, the first row is the control input. It is a quote from the book *A Tale of Two Cities*. You will find that the augmented effects are easier to spot by comparing the control text with the text in the datasets. In addition, the control text is needed to compare the differences between augmentation techniques.

Pluto found the injection of names on *row #1*, such as **Punta** (a believable Spanish writer name) and **Poydras**, as actual names and plausible additions to this celebrity movie review context. It was not factual in the movie, but it is acceptable for text augmentation for movie sentiment prediction.

On *row #2*, the words **blending**, **dangerous**, and **original 1960s** add flare to the movie description without altering the intent of the spy movie's description.

On *row #3*, the addition of names, such as **Kent of Cabus** (Kent from a village in English named Cabus), **Rangjung** (a village in Bhutan, served as a possible hero name), and **Elizabeth** (as the villain) in the comic Green Arrow movie description is 100% plausible plot for our superhero.

Overall, Pluto is flabbergasted by the **Word2Vec** ML model. The word and name injections are contexts that are appropriate as if a human writer were creating them. However, the control text from Dickens is funny to read, and it is not ML's fault. The system does not know that the book was written in the 1800s and has only the first few lines of the text to go off. The movie review is a complete thought, while the control text is a tiny fragment of the whole.

Pluto runs a similar command on the real-world Twitter data, as follows:

```
# augment using word2vec
pluto.print_aug_ai_word2vec(pluto.df_twitter_data,
  col_dest='clean_tweet',
  action='insert',
  aug_name='Word2Vec-GoogleNews Word Embedding Augment')
```

The output is as follows:

	Word2Vec-GoogleNews Word Embedding Augment	Original
0	It was the best of times. Dale It was the worst of Scrappage times. It ingredient was the age of Home wisdom. It was Esam the age of foolishness. It FREE was Minister the epoch Karl of belief. PTI It was the batshit epoch of incredulity.	It was the best of times. It was the worst of times. It was the age of wisdom. It was the age of foolishness. It was the epoch of belief. It was the epoch of incredulity.
1	@ Lynda user find your MANCHESTER place Schleiss how NORTH to outsma # stress # zensei in # miamishores with Olivant your # Mohammed family in # Borneo min	@user find your place how to outsma #stress #zensei in #miamishores with your #family in #min
2	uh oh, warranty lattice on @ Bendetson user humanoid is Elizabeth apparently exactly Puks 70 years. SOUTH ing Pesticide himself. # daisydaisy	uh oh, warranty on @user humanoid is apparently exactly 70 years. ing himself. #daisydaisy
3	not anymore. Amphlett im Coatesville getting maccas Reigns for dinner	not anymore. im getting maccas for dinner

Figure 6.5 – Word2Vec using insert mode on the Twitter data

Since tweets are like random thoughts written without forethoughts or editing, in *Figure 6.4*, the **Word2Vec** injections are like a bored high school student doing homework while playing a computer game. Pluto can't judge if the altered text is plausible or not. Would it increase or decrease the AI prediction accuracy for sentiment analysis?

For Dickens' control text, Pluto flinched. It was dreadful, but he promised the AI would be better in the later model when using transformers and generative AI.

Now that we've looked at **insert** mode, let's see how the **Word2Vec** model performs in **substitute** mode.

Substitute

Substitute mode replaces words and then adds words to the sentence. Pluto applies the **Word2Vec** model using **substitute** mode to the Netflix data like so:

```
# augmenting using word2vec
pluto.print_aug_ai_word2vec(pluto.df_netflix_data,
    col_dest='description',
    action='substitute',
    aug_name='Word2Vec-GoogleNews Word Embedding Augment')
```

The output is as follows:

	Word2Vec-GoogleNews Word Embedding Augment	Original
0	Usually was ultimately best of times. It was the worst of times. It was those age of wisdom. It wasthe certainly Matures_ages_##-## of foolishness. It was this paradigm_shifts of belief. It was on Sokoto_Caliphate of incredulity.	It was the best of times. It was the worst of times. It was the age of wisdom. It was the age of foolishness. It was the epoch of belief. It was the epoch of incredulity.
1	Musicloving baby Johny goes before dancefilled zany_adventures withtheir consoling_pat family, friends and adorable critters early LiquidViagra colorful, live-action/CGI preschooler series.	Musicloving baby Johny goes on dancefilled adventures with his family, friends and adorable critters in this colorful, animated preschooler series.
2	An intrepid troopers inspector forced into a desk plum_assignment wants wrest matters into his own finger when ruthlessly_efficient offenders should released on a technicality.	An intrepid police inspector forced into a desk job must take matters into his own hands when ruthless criminals are released on a technicality.
3	Thus this Scorcese decaffeination of the sparkling stage performance, five spirited souls take on whole dark threat growing when clandestine Skull Castle.	In this cinematic distillation of the electrifying stage performance, seven spirited souls take on the dark threat growing in shadowy Skull Castle.

Figure 6.6 – Word2Vec using substitute mode on the Netflix data

In *Figure 6.5*, *row #0* is the control text, and on *row #1*, **zany adventure** is suitable for a kid adventure movie, but **liquid viagra** is definitely off the mark.

On *row #2*, replacing **police** with **troopers**, **job** with **plum assignment**, **wrest** with **must take**, **figure** with **hand**, and **offenders** with **criminals** are suitable in the police movie. Thus, the **Word2Vec** model did a proper augmentation job.

On *row #3*, replacing **cinematic distillation** with **Scorcese decaffeination** is an intriguing choice worthy of a human writer. Changing **electrifying** to **sparkling** is clever because electricity can spark. Substituting **shadowy** with **clandestine** is a good choice, but switching **seven** with **five** is unnecessary.

Once again, the **Word2Vec** model could have done better for the control text.

Pluto does the same to the Twitter data with the following command:

```
# augmenting using word2vec
pluto.print_aug_ai_word2vec(pluto.df_twitter_data,
    col_dest='clean_tweet',
    action='substitute',
    aug_name='Word2Vec-GoogleNews Word Embedding Augment')
```

The output is as follows:

	Word2Vec-GoogleNews Word Embedding Augment	Original
0	Such originally the best of times. It was the worst of times. Unfortunately went the age of wisdom. It was the mid_##s of foolishness. It wasa by aeon of belief. Probably itwas the epoch of incredulity.	It was the best of times. It was the worst of times. It was the age of wisdom. It was the age of foolishness. It was the epoch of belief. It was the epoch of incredulity.
1	your debate style 1 refamiliarize up facts six use especially naff no programmed to line four mute Ulrick_dentist # weak # patheticreally	your debate style 1 make up facts 2 use some lame not programmed to line 3 mute them #weak #patheticreally
2	@ logins @ user how tomorrow_Camardelle you not know just chancery lane currently? # GPS pull E_mail_heyjen@phillynews.com to a lake? # poorservice # reacted_furiously	@user @user how do you not know where chancery lane is? #satnav send you to a lake? #poorservice #infuriated
3	Whose people would have all died of Spanish_Armada consumersâ_€_™ thrid winter if the first TOYAKO_Japan_Reuters reall shown mountainside_Kugler a cute.	your people would have all died of scurvy their first winter if the first nations hadnt shown them a cute.

Figure 6.7 – Word2Vec using substitute mode on the Twitter data

In *Figure 6.6*, the tweets are chaotic, and many are incomplete thoughts. The **Word2Vec** model does its best, and Pluto doesn't think a human can do better.

The next technique we'll look at is **BERT**, which uses the transformer model and generative AI.

BERT

BERT is a Google transformer model trained on a massive corpus. The result is a near-perfect human-quality output. BERT and many other transformer models were made available and easily accessible on *HuggingFace* to the public starting around mid-August 2022.

The key code lines for the `print_aug_ai_bert()` function are as follows:

```
# code snippet for print_aug_id_bert()
model_path='bert-base-uncased'
aug_func = nlpaug.augmenter.word.ContextualWordEmbsAug(
  action=action,
  model_path=model_path)
```

The full function can be found in the Python Notebook. Pluto feeds in the NLP Netflix data using `insert` mode with the following command:

```
# Augmenting using BERT
pluto.print_aug_ai_bert(pluto.df_netflix_data,
  col_dest='description',
  action='insert',
  aug_name='BERT Embedding Insert Augment')
```

The result is as follows:

	BERT Embedding Insert Augment	Original
0	it was the old best of times. it was the whole worst of times. it once was the next age advance of wisdom. then it was the next age advance of foolishness. it was the epoch stage of belief. it was the epoch of extreme incredulity.	It was the best of times. It was the worst of times. It was the age of wisdom. It was the age of foolishness. It was the epoch of belief. It was the epoch of incredulity.
1	afterwards a teacher turns to a financially dubious man hoping to raise money for her daughters with lifesaving knee surgery, setting a crisis story that is also unexpectedly catapulted to play a national film stage.	A teacher turns to a dubious man to raise money for her daughters lifesaving surgery, a crisis that is unexpectedly catapulted to a national stage.
2	a war criminal whilst in hiding begins to deeply suspect that the unknown maid, perhaps his only confidant and unlikely contact with the outside russian world, also may be hiding quite something from herself.	A war criminal in hiding begins to suspect that the maid, his only confidant and contact with the outside world, may be hiding something herself.
3	arriving in february 1895 mumbai, spirited little indian lad called shiv talpade defies british rule by and despite sparse funding mathematics and physics to build the two worlds first functioning airplane.	In 1895 Mumbai, spirited Indian lad Shiv Talpade defies British rule and sparse funding and physics to build the worlds first airplane.

Figure 6.8 – BERT using insert mode on the Netflix data

In *Figure 6.7*, Pluto immediately recognizes the improvement over the **Word2Vec** model. In the control text, *row #0*, the injection of words is acceptable. It lacks the elegance of the prose, but if you must add words to the text, it could pass as a human writer.

In *row #1*, the added phrases are spot on, such as **afterward**, **financial dubious**, **knee surgery**, and **to play a national film stage**.

In *row #2*, the augmented phrases are at human writer quality, such as **whilst in hiding**, **deeply suspect**, **unknown maid**, **perhaps his only**, **outside Russian world**, and **maybe hiding quite something**.

In *row #3*, Pluto is impressed with the results, such as **arriving in February**, **little Indian lad**, **despite sparse funding**, **funding mathematics and physics**, and **first functioning airplane**.

> **Fun fact**
>
> Are you as amazed as Pluto regarding the BERT model's output? It is like BERT is a real person, not an ANN.

Please rerun the wrapper function to see BERT's augmentation on other movie reviews. The more you read, the more you will appreciate the advanced breakthrough in using the **transformer** model. It is the foundation of generative AI.

Next, Pluto feeds the Twitter data into the BERT model with insert mode with the following command:

```
# augmenting using BERT
pluto.print_aug_ai_bert(pluto.df_twitter_data,
  col_dest='clean_tweet',
  action='insert',
  aug_name='BERT Embedding Insert Augment')
```

The result is as follows:

	BERT Embedding Insert Augment	Original
0	sometimes it was the best of times. it even was still the worst good of times. it was once the age old of learned wisdom. before it was the age age of foolishness. it was the epoch of belief. it was the epoch times of incredulity.	It was the best of times. It was the worst of times. It was the age of wisdom. It was the age of foolishness. It was the epoch of belief. It was the epoch of incredulity.
1	just watched @ user wish you i ddnt no they wrecked but it is not looking foward to dream next season # bachelor season4 # celebrity netflex # celebrity worstthingever # the notfunny # damn shit	just watched @user wish i ddnt they wrecked it not looking foward to next season #season4 #netflex #worstthingever #notfunny #shit
2	best of your luck @ blog user @ www user	best of luck @user @user
3	@ user its quite clear what yall had made no good reason to ever hand the program keys automatically over quickly to a malicious goon. but yet...	@user its clear yall had no good reason to hand the keys over to a goon. but yet...

Figure 6.9 – BERT using insert mode on the Twitter data

> **Fun fact**
>
> In *Figure 6.8*, the BERT model gives another version of Dicken's control text. There is a new rendition every time Pluto runs the wrapper function. The possibilities are endless. Pluto must have run the wrapper functions over 50 times. Not once did he notice the same result.

Pluto discovered that there is better NLP data to study than tweets, but they represent the real world, so it is worth continuing to use them. As Pluto repeatedly rerun the wrapper function, he preferred the BERT augmented version over the original tweets because inserting text made it easier to read.

When switching to **substitute** mode, the output from BERT is better than average human writers.

Substitute

Next, Pluto feeds the Netflix data to BERT in **substitute** mode using the following command:

```
# augmenting using BERT
pluto.print_aug_ai_bert(pluto.df_netflix_data,
    col_dest='description',
    action='substitute',
    aug_name='BERT Embedding Substitute Augment')
```

The result is as follows:

	BERT Embedding Substitute Augment	Original
0	it was the best for all. it was the lowest of times. it gave the age of wisdom. death was the age of love. it was the epoch in belief. it was an time of life.	It was the best of times. It was the worst of times. It was the age of wisdom. It was the age of foolishness. It was the epoch of belief. It was the epoch of incredulity.
1	with their relationship rushed to the final on christmas day, a few cannot decide if theyre really gay and hand in a towel on their relationship.	With their divorce set to become final on Christmas day, a couple must decide if theyre really ready to throw in the towel on their relationship.
2	in a live setting, learning... cocaine provide students, athletes, coders and others one way to learn more faster but better. but for what cost?	In a hypercompetitive world, drugs like Adderall offer students, athletes, coders and others a way to do more faster and better. But at what cost?
3	follow and life of superspy danger mole and his bumbling villain, penfold, while they return to the country performing their supposed evil plots.	Follow the adventures of superspy Danger Mouse and his bumbling sidekick, Penfold, as they jet around the world foiling their enemies evil plots.

Figure 6.10 – BERT using substitute mode on the Netflix data

In *Figure 6.9*, for the control text, *row #0*, BERT replaced **it was the age of foolishness** with **death was the age of love**.

Fun fact

Full stop. Pluto's mind is being blown. Even Pluto's human companion is speechless. Pluto expects a transformer model such as BERT to be good, but philosophical thoughts or poetry are on another level. Now, are you impressed with BERT?

The rest of the movie review augmentation, shown are *rows #1, #2,* and *#3,* is flawless. The augmented words match the movie genre and context. It is like BERT understands the movie's meaning, but this isn't true. The BERT model is no more sentient than a toaster. However, BERT can mimic a human writer well.

One interesting note is that in *row #1,* in the movie description about a couple's relationship, BERT uses the word *gay,* which was discussed in the previous chapter about data biases. This is because *gay* is a perfectly nice word for lighthearted and carefree, but in a modern context, *gay* is associated with a person's homosexual orientation, especially of a man.

Once again, Pluto encourages you to rerun the wrapper function repeatedly on the Python Notebook. You will appreciate it beyond the technical achievement and think that BERT has a personality.

Pluto does the same for the Twitter data with the following command:

```
# augmenting using BERT
pluto.print_aug_ai_bert(pluto.df_twitter_data,
  col_dest='clean_tweet',
  action='substitute',
  aug_name='BERT Embedding Substitute Augment')
```

The result is as follows:

	BERT Embedding Substitute Augment	Original
0	it had the best of everything. it saw the end of times. it was the age of wisdom. history included the age of foolishness. it was the age of youth. i was the epoch to incredulity.	It was the best of times. It was the worst of times. It was the age of wisdom. It was the age of foolishness. It was the epoch of belief. It was the epoch of incredulity.
1	most i seemed to be doing down here was getting drunk at nappers or home in bed	all i wanted to be doing right now was getting drunk at nappers but im in bed
2	so all because these cool touch 6 message things probably exist if that person youre messaging has ios 7, and no type i know doesssssss	so all of the cool ios 10 message things only work if the person youre messaging has ios 10, and no one i know doesssssss
3	always keeping my suppo with neighbors! # 1 # y # b	always like to suppo our peeps! #ameliacaruso #a #suppolocalaists

Figure 6.11 – BERT using substitute mode on the Twitter data

As Pluto repeatedly ran the wrapper function on the Python Notebook, in *Figure 6.10*, he found that the augmented tweets were more accessible to read than the original text.

For the control text, *row #0*, Pluto found that having the augmented text **it was the age of youth**, replace the original text of **it was the epoch of belief** profoundly appropriate. It fits into the context and style of Mr. Dickens's book.

> **Fun challenge**
>
> This challenge is a thought experiment. BERT is built on an ANN algorithm. It does not contain grammar rules, such as nouns and verbs for constructing sentences. With no grammar rules, how does it write English so well? Hint: think about patterns. BERT is trained on a massive corpus. The number of words and sentences is so large that it was impossible to conceive 5 years ago. A few, if any, know how neural network algorithms learn. It is not complex math. It is gradient descent and matrix multiplication nudging billions of nodes (or neurons), but how does that collection of nodes write English convincingly?

Pluto can spend days talking about BERT, but let's move forward with **RoBERTa (Roberta)**. It sounds like a female version of BERT.

RoBERTa

RoBERTa is an optimized algorithm for self-supervising BERT. While Google created BERT, Meta AI (or Facebook) developed RoBERTa.

Pluto feeds the Netflix data to RoBERTa in insert mode with the following command:

```
# augmenting using Roberta
pluto.print_aug_ai_bert(pluto.df_netflix_data,
  col_dest='description',
  model_path='roberta-base',
  action='insert',
  aug_name='Roberta Embedding Insert Augment')
```

The result is as follows:

	Roberta Embedding Insert Augment	Original
0	It certainly was the very best of times. It was perhaps the worst of all times. It was the vast age of wisdom. It was not the age of foolishness. It it was the epoch short of belief. It was the supreme epoch of utter incredulity.	It was the best of times. It was the worst of times. It was the age of wisdom. It was the age of foolishness. It was the epoch of belief. It was the epoch of incredulity.
1	The global spy game clearly is today a serious business, and and perhaps throughout all history, the tools tools and technologies developed specifically for it have mattered as much as ever the spies themselves.	The spy game is a serious business, and throughout history, the tools and technologies developed for it have mattered as much as the spies themselves.
2	A devoted young grandsons determined mission to reunite the his ailing Irish grandmother with in her Taiwanese ancestral home turns into some a really complicated, sometimes comic crossborder affair.	A devoted grandsons mission to reunite his ailing grandmother with her ancestral home turns into a complicated, comic crossborder affair.
3	In a series full of inspiring homemade home makeovers, internationally worldrenowned tidying techniques expert Ellen Marie Marie Kondo helps many clients clear up out the clutter and choose true joy.	In a series of inspiring home makeovers, worldrenowned tidying expert Marie Kondo helps clients clear out the clutter and choose joy.

Figure 6.12 – RoBERTa using insert mode on the Netflix data

The output in *Figure 6.11* is similar to the output from **BERT**, which is impressive. The words are not randomly inserted in the sentence. They expressed a possible interpretation and gave the impression that **RoBERTa** understood the meaning of the words. This level of technical achievement was not feasible 1 year ago, and **RoBERTa** was only made available a few months ago.

Pluto ran the wrapper function repeatedly and never tired of reading the result. He does the same for the Twitter data with the following command:

```
# augmenting using Roberta
pluto.print_aug_ai_bert(pluto.df_twitter_data,
    col_dest='clean_tweet',
    model_path='roberta-base',
    action='insert',
    aug_name='Roberta Embedding Insert Augment')
```

The result is as follows:

	Roberta Embedding Insert Augment	Original
0	It returned was once the best of times. It repeated was the worst of times. It once was again the age full of no wisdom. It was the age of foolishness.. It again was the epoch of belief. It also was the epoch of incredulity.	It was the best of times. It was the worst of times. It was the age of wisdom. It was the age of foolishness. It was the epoch of belief. It was the epoch of incredulity.
1	#christinagrimmie	#usa @ #gunlaws @ #stop # #violence # #rip #news #shot she he was only around 22...	#christinagrimmie #usa #gunlaws #stop #violence #rip #news #shot she was only 22...
2	its been only barely been a half day i already miss them!!	its only been a day i already miss them!!
3	@user weddings that are an impoant nonsense because really they merely celebrate life joy and new possibility. anne hathaway #wedding #love	@user weddings are impoant because they celebrate life and possibility. anne hathaway #wedding #love

Figure 6.13 – RoBERTa using insert mode on the Twitter data

Pluto can't turn lead into gold, and RoBERTa can't turn tweets, as shown in *Figure 6.12*, that contain misspellings and incomplete thoughts into coherent sentences. Nevertheless, RoBERTa is one of the best choices for augmenting real-world tweets.

Next, Pluto will try **RoBERTa** with **substitute** mode.

Substitute

In substitute mode, RoBERTa will replace words or phrases with uncanny accuracy matching the context and writing style.

Pluto drops the Netflix data into the RoBERTa model in substitute mode using the following command:

```
# augmenting using Roberta
pluto.print_aug_ai_bert(pluto.df_netflix_data,
  col_dest='description',
  model_path='roberta-base',
  action='substitute',
  aug_name='Roberta Embedding Substitute Augment')
```

The output is as follows:

	Roberta Embedding Substitute Augment	Original
0	It was the best at periods. It was the worst of times. It appeared the age before wisdom. It seemed it age of joy. It was that epoch of belief. It was its epoch of innocence.	It was the best of times. It was the worst of times. It was the age of wisdom. It was the age of foolishness. It was the epoch of belief. It was the epoch of incredulity.
1	A disastrous rescue strands Alex the lion and two companions in Africa. However, Alex discovers Felix had special romantic chemistry with the locals.	A botched rescue strands Alex the lion and his companions in Africa. Sadly, Alex discovers he has little in common with the locals.
2	After causing the very end of mankind seven years earlier, genius theorist Arthur Walker must now fight every ecological disaster he helped created.	After causing the near extinction of mankind seven years ago, genius scientist Leon Lau must now fight the ecological disaster he unwittingly created.
3	It lit sparked rap and hiphop and ignited a proliferation of drugs entangled in racial injustice. Few believe Americas deeper histories with weed.	It lit up jazz and hiphop and ignited a war on drugs steeped in racial injustice. Experts explore Americas complicated relationship with weed.

Figure 6.14 – RoBERTa using substitute mode on the Netflix data

No matter how often Pluto executes the wrapper function, he continues to be astonished by the output RoBERTa provides in *Figure 6.13*. For example, in *row #1*, she changed the phrase **Alex discovers he has little in common with the local** to **Alex discovers Flix had special romantic chemistry with the local**. RoBERTa has quite the imagination. Is that what humans do when we write?

Pluto does the same with the Twitter data using the following command:

```
pluto.print_aug_ai_bert(pluto.df_twitter_data,
    col_dest='clean_tweet',
    model_path='roberta-base',
    action='substitute',
    aug_name='Roberta Embedding Substitute Augment')
```

The result is as follows:

	Roberta Embedding Substitute Augment	Original
0	It was the dawn of times. It was it brightest among times. It was the age of wisdom. It woke the age of foolishness. It preached an curse at arrogance. It was the epoch of incredulity.	It was the best of times. It was the worst of times. It was the age of wisdom. It was the age of foolishness. It was the epoch of belief. It was the epoch of incredulity.
1	#dancerproblems computer simulation that could t climb the surrounding range of mountains. Have reached the leakage	#dancerproblems rooster simulation i want to climb the vast expanse of mountains. it reached the leakage
2	#hamzayusuf might be given a #ris # by twitter and hilton fox	#hamzayusuf apologists be like ... #ris #jetlagged @ doubletree by hilton london
3	@user Find us page for more fun random gems at #inspirationalquotes	@user like this page for more badass motivational quotes gt #inspirationalquotes

Figure 6.15 – RoBERTa using substitute mode on the Twitter data

As shown in *Figure 6.14*, text augmentation does not have to be boring or clinical. Using transformer models such as BERT and RoBERTa, augmentations are fun and full of wonders. For example, in the control text, on *row #0*, RoBERTa wrote, **It preached a curse at arrogance**, replacing **It was an epoch of belief**.

Fun fact

Pluto's human companion has to ponder a long time to conclude that the augmented text does mean the same as the original text in *Figure 6.14*, the control text. It is easy to be fooled that RoBERTa has a conscience. We pair intelligence with consciousness, meaning if you have intelligence, you must be self-aware or vice versa, but we know that is not true. For example, a career politician is self-aware. He talks about himself all the time, but is he intelligent?

Continuing to use the latest powerful ML models, Pluto will take a different path to text augmentation by using the **back translation** technique.

Back translation

Back translation is a new concept in text augmentation because it was not possible 2 years ago. ML NLP existed earlier, with Google Translate leading the charge. Still, only a few data scientists could access the large language model using a transformer and the powerful servers required for language translation.

The technique for text augmentation is to translate into another language and back to the original language. In doing so, the result will be an augmented version of the original. No language translation is perfect. Hence, the extended version will be slightly different from the original text. For example, the original text is in English. Using a powerful NLP model, we translated it into German and back to English again. The translated English text will be different from the original English text.

Compared to **Word2Vec**, **BERT**, and **RoBERTa**, the back translation method could be more robust. This is because translating back to the original text gives the same result the second or third time. In other words, other methods' output results in thousands of variations, while back translations have two or three augmented versions.

Pluto found two NLP pre-trained translation models from Facebook, or Meta AI, that were made available on the *HuggingFace* site. They are for English to German (Deutsch) and English to Russian (Русский). There are dozens more, but two are sufficient to demonstrate the technique. Let's start with German.

German (Deutsch)

The `print_aug_ai_back_translation()` method follows the same structure as any other wrapper function. It looks deceptively simple with five lines of code, but it has truly complex theories and coding techniques. It reminds Pluto of a famous quote by Sir Isaac Newton: *"If I have seen further, it is by standing on the shoulders of giants."*

The key code lines are as follows:

```
# code snippet for back translation
from_model_name='facebook/wmt19-en-de'
to_model_name='facebook/wmt19-de-en'
aug_func = nlpaug.augmenter.word.BackTranslationAug(
  from_model_name=from_model_name,
  to_model_name=to_model_name)
```

The full function can be found in the Python Notebook. Pluto feeds in the Netflix data using the following command:

```
# augmenting using back translation
pluto.print_aug_ai_back_translation(pluto.df_netflix_data,
```

```
col_dest='description',
from_model_name='facebook/wmt19-en-de',
to_model_name='facebook/wmt19-de-en',
aug_name='FaceBook Back Translation: English <-> German Augment')
```

The result is as follows:

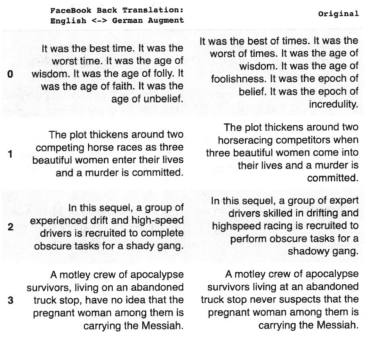

	FaceBook Back Translation: English <-> German Augment	Original
0	It was the best time. It was the worst time. It was the age of wisdom. It was the age of folly. It was the age of faith. It was the age of unbelief.	It was the best of times. It was the worst of times. It was the age of wisdom. It was the age of foolishness. It was the epoch of belief. It was the epoch of incredulity.
1	The plot thickens around two competing horse races as three beautiful women enter their lives and a murder is committed.	The plot thickens around two horseracing competitors when three beautiful women come into their lives and a murder is committed.
2	In this sequel, a group of experienced drift and high-speed drivers is recruited to complete obscure tasks for a shady gang.	In this sequel, a group of expert drivers skilled in drifting and highspeed racing is recruited to perform obscure tasks for a shadowy gang.
3	A motley crew of apocalypse survivors, living on an abandoned truck stop, have no idea that the pregnant woman among them is carrying the Messiah.	A motley crew of apocalypse survivors living at an abandoned truck stop never suspects that the pregnant woman among them is carrying the Messiah.

Figure 6.16 – Back translation, German on Netflix data

Fun fact

The output in *Figure 6.15* is anticlimactic because it reads similarly to the original text, but the technical achievement is mind-blowing. First, you need an expert to translate from English to German. It is a challenging task for a human to learn. Second, you must translate back to English with no errors. The difference in choosing similar words is expressing the phrase. Maybe 5% of the world's population can do this task. For a machine to do it 24 hours a day, 7 days a week, and maintain the same accuracy level is miraculous. No human can match this level.

The output in *Figure 6.15* gives an almost perfect English to German and back translation. Pluto does the same with the Twitter data using the following command:

```
# augmenting using back translation
pluto.print_aug_ai_back_translation(pluto.df_twitter_data,
```

```
col_dest='clean_tweet',
from_model_name='facebook/wmt19-en-de',
to_model_name='facebook/wmt19-de-en',
aug_name='FaceBook Back Translation: English <-> German Augment')
```

The output is as follows:

	FaceBook Back Translation: English <-> German Augment	Original
0	It was the best time. It was the worst time. It was the age of wisdom. It was the age of folly. It was the age of faith. It was the age of unbelief.	It was the best of times. It was the worst of times. It was the age of wisdom. It was the age of foolishness. It was the epoch of belief. It was the epoch of incredulity.
1	@ user posted the first time a year. still relevant, but not for me!	@user first posted this a year ago. still relevant, though not for me! heres my #thoughtfooday #valued
2	fathersday # fathers # day, # dad # # skinny # and # single buy things about h	fathersday #fathers #day, #dad # #skinny #and #single buy things about h
3	latepost # dinner # friends # breakfasting # surabaya # instagood # red # bff # l4l # instapict	latepost #dinner #friends #breakfasting #surabaya #instagood #red #bff #l4l #instapict

Figure 6.17 – Back translation, German on Twitter data

In *Figure 6.16*, translating nonsensible tweets into German and back is harder for humans because our minds get tired more quickly and give up. Only a machine can do this work around the clock. The control text translations into German and back are acceptable.

Translation to Russian and back would yield similar results. Let's take a look.

Russian (Русский)

Pluto chose to repeat the same back translation technique with English to Russian and back because he is curious to see if choosing a non-Romance family language would affect the augmentation results differently.

Using the same `print_aug_ai_back_translation()` function, Pluto defines the Russian translation Facebook model as follows:

```
# code snippet for back translation to Russian
from_model_name='facebook/wmt19-en-ru'
to_model_name='facebook/wmt19-ru-en'
```

The full function code can be found in the Python Notebook. Pluto feeds the Netflix data to the wrapper function as follows:

```
# augmenting using back translation
pluto.print_aug_ai_back_translation(pluto.df_netflix_data,
    col_dest='description',
    from_model_name='facebook/wmt19-en-ru',
    to_model_name='facebook/wmt19-ru-en',
    aug_name='FaceBook Back Translation: English <-> Russian Augment')
```

The result is as follows:

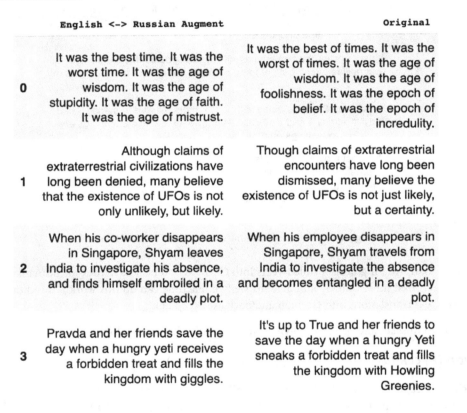

	English <-> Russian Augment	Original
0	It was the best time. It was the worst time. It was the age of wisdom. It was the age of stupidity. It was the age of faith. It was the age of mistrust.	It was the best of times. It was the worst of times. It was the age of wisdom. It was the age of foolishness. It was the epoch of belief. It was the epoch of incredulity.
1	Although claims of extraterrestrial civilizations have long been denied, many believe that the existence of UFOs is not only unlikely, but likely.	Though claims of extraterrestrial encounters have long been dismissed, many believe the existence of UFOs is not just likely, but a certainty.
2	When his co-worker disappears in Singapore, Shyam leaves India to investigate his absence, and finds himself embroiled in a deadly plot.	When his employee disappears in Singapore, Shyam travels from India to investigate the absence and becomes entangled in a deadly plot.
3	Pravda and her friends save the day when a hungry yeti receives a forbidden treat and fills the kingdom with giggles.	It's up to True and her friends to save the day when a hungry Yeti sneaks a forbidden treat and fills the kingdom with Howling Greenies.

Figure 6.18 – Back translation, Russian on Netflix data

Remarkably, in *Figure 6.17*, the NLP **T5** model translates a Romance family language (English) into an East Slavic language (Russian) and back with almost perfect accuracy. The grammar rules, sentence structures, alphabets, histories, cultures, and languages are different, yet a machine can do the task without awareness.

Tweets are not perfect for testing, but not all projects are logical. Pluto had worked on real-world NLP projects that were ill-conceived. The command for feeding Twitter data to the wrapper function is as follows:

```
# augmenting using back translation
pluto.print_aug_ai_back_translation(pluto.df_twitter_data,
    col_dest='clean_tweet',
    from_model_name='facebook/wmt19-en-ru',
    to_model_name='facebook/wmt19-ru-en',
    aug_name='FaceBook Back Translation: English <-> Russian Augment')
```

The result is as follows:

	FaceBook Back Translation: English <-> Russian Augment	Original
0	It was the best time. It was the worst time. It was the age of wisdom. It was the age of stupidity. It was the age of faith. It was the age of mistrust.	It was the best of times. It was the worst of times. It was the age of wisdom. It was the age of foolishness. It was the epoch of belief. It was the epoch of incredulity.
1	are you # black amp feel like the are stopping on you? # retweet # tampa # Miami	are you #black amp feel like the are stomping on you? #retweet #tampa #miami
2	I'm kind of adding to this # avocado on # sourdough and # coffee # Friday # breakfast # eatclean	im kind of addicted to this #avocado on #sourdough and #coffee #friday #breakfast #eatclean
3	i am grateful for my friends. # grateful # positive	i am thankful for my friends. #thankful #positive

Figure 6.19 – Back translation, Russian on Twitter data

If Russians don't understand tweets, then who else can? Reading the control text in *Figure 6.18*, Pluto can tell the translations are correct. Since some tweets are short, the translations to Russian and back are perfect.

> **Fun challenge**
>
> This challenge is a thought experiment. Can you use the same techniques to augment the German language? Or can you string several back translations together – for example, from English to German to Russian and back to English?

The **back translation**, **RoBERTa**, **BERT**, and **Word2Vec** NLP ML models are the state of the art for text augmentation. The next level is sentence augmentation using summarization and the Sequential and Sometimes techniques.

Sentence augmentation

The sentence **flow** level uses a combination of word augmentation methods. But before that, Pluto will use the **T5** NLP model to generate a text summary. The **summarization** technique is one of the novel concepts made possible recently. It takes a page, an article, or even a book and generates a summary to be used in the NLP text augmentation model.

Summary technique

For text augmentation, the **summary** technique may bring a few different versions for training. However, suppose Pluto combines the **flow** and **summary** techniques, such as by feeding the synopsis text, instead of the original text, to the **flow** technique. In that case, it will yield many new original texts for training.

> **Fun fact**
>
> Pluto pioneered the **summary-to-flow** concept for text augmentation. He had done a preliminary search on the web and scholarly publications, but he needs help finding a reference to the summary-to-flow technique. If none are found, then Pluto is the first to implement the summary-to-flow strategy.

Pluto will not use the Netflix movie description or Twitter tweets for the summary method. This is because they are too short to showcase the power of the T5 NLP model. Instead, Pluto will use the first page of the following books mentioned in the *Real-world NLP data* section:

- *Tale of Two Cities* by Dickens
- *Moby Dick* by Melville
- *Alice in Wonderland* by Carroll

Once again, the books are in the public domain, as defined in Project Gutenberg.

In addition, Pluto will use the first page of this chapter because you have read this book, but you may not have read those three classic books.

The key code line for the `print_aug_ai_t5()` wrapper function is as follows:

```
aug_func = nlpaug.augmenter.sentence.AbstSummAug(
    model_path='t5-base')
```

Pluto is playing a guessing game with you. First, he will list the four command lines to generate the four summaries, but he will shuffle the output. Thus, you have to guess which summary belongs to which book. Once again, you will be amazed at the quality output of the **T5** NLP model. It is comparable to human writers.

The profound implication is that you or Pluto can auto-generate summaries of books, papers, documents, articles, and posts with a few lines of Python code. This task was deemed impossible a few years ago.

Fun challenge

Here is a thought experiment. Can you be an expert in German laws without speaking German? It was impossible a year ago because the ML breakthrough wasn't available, but today, you can use the code in the Python Notebook as the base to translate all German law books.

The four commands to get a summary of the first page of the four books we'll be looking at are as follows:

```
# Alice in Wonderland
pluto.df_t5_carroll = pluto.print_aug_ai_t5(
    pluto.orig_carroll_page,
    bsize=1)
# Tale of Two Cities
pluto.df_t5_dickens = pluto.print_aug_ai_t5(
    pluto.orig_dickens_page,
    bsize=1)
# Moby Dick
pluto.df_t5_melville = pluto.print_aug_ai_t5(
    pluto.orig_melville_page,
    bsize=1)
# This chapter first page
pluto.df_t5_self = pluto.print_aug_ai_t5(
    pluto.orig_self,
    bsize=1)
```

The shuffled results are as follows:

T5_summary

0	david rothkopf: spiritual revelations were conceded to England at favoured time. he says even the Cock-lane ghost had been laid only a round dozen of years. the spirits of this very year last past rapped out theirs, he writes.

Figure 6.20 – Summary T5 NLP engine – 1

The second output is as follows:

T5_summary

0   Ishmael is a sailor who has a way of driving off the spleen and regulating the circulation. he says it is his substitute for pistol and ball to get to sea as soon as he can. almost all men in their degree, some time or other, cherish very nearly the same feelings.

Figure 6.21 – Summary T5 NLP engine – 2

The third output is as follows:

T5_summary

0   text augmentation with machine learning (ML) is an advanced technique. we used a pre-trained ML model to create additional training NLP data.

Figure 6.22 – Summary T5 NLP engine – 3

The fourth output is as follows:

T5_summary

0   a white rabbit with pink eyes ran close by. the rabbit took a watch out of its waistcoat-pocket, and then hurried on. she ran across the field after it, and was just in time to see it pop down a rabbit-hole.

Figure 6.23 – Summary T5 NLP engine – 4

> **Fun challenge**
> Can you match the summarized output with the book? The T5 engine is not a generative AI engine like OpenAI GPT3, GPT4, or Google Bard. Still, the summary is very accurate.

The *Tale of Two Cities* book, shown in *Figure 6.19*, is a relatively hard book to read, and Pluto thinks that it is funny that *David Rothkopf*, a contemporary political commentator, is associated with Dickens' book. The first page does talk about the **congress of British subjects in America**. Thus, the Mr. Rothkopf association is a good guess. Maybe Pluto should feed the first 10 pages of the chapter into the **T5** NLP engine and see the summary.

The *Moby Dick* first-page summary is spot on, as shown in *Figure 6.20*. It could pass as a human writer, and the first word is **Ishmael**. Pluto wishes that the **T5** NLP model was available during Pluto's early days in school.

Pluto's human companion is delighted to admit that the summary of this chapter's first page is more precise and easier to read, as shown in *Figure 6.21*. Maybe the **T5** NLP engine should co-write this book with Pluto so that his companion can enjoy chasing squirrels on a sunny afternoon.

The *Alice in Wonderland* first-page summary is perfect, as shown in *Figure 6.22*. The **T5** NLP engine captures the assent of the opening page flawlessly. As a bonus, Pluto only inputted the first five sentences. The output is as follows:

T5_summary

0   white rabbit with pink eyes runs close by a daisy-chain maker. the daisies are made from a variety of flowers.

Figure 6.24 – Summary T5 NLP Engine – the first five lines of Alice in Wonderland

In *Figure 6.23*, how does **T5** know that the white rabbit is essential to the story? The rabbit only appears in the last sentence in the input text, and referring to Alice as the daisy-chain maker is delightful.

The next step in sentence augmentation is to feed these summaries to the **flow** methods.

Summary-to-flow technique

The Sequential method in the flow technique applies a list of augmentation in successive order. Pluto creates two text augmentation methods, as follows:

```
# augment using uloVe
pluto.ai_aug_glove = nlpaug.augmenter.word.WordEmbsAug(
    model_type='glove', model_path='glove.6B.300d.txt',
    action="substitute")
# augment using BERT
pluto.ai_aug_bert = nlpaug.augmenter.word.ContextualWordEmbsAug(
  model_path='bert-base-uncased',
  action='substitute',
  top_k=20)
```

The first uses the **Word2Vec** with the **GloVe** model, while the second employs the **BERT** NLP engine. The `print_aug_ai_sequential()` wrapper function uses the augmentation list with the key code lines, as follows:

```
# augment using sequential
aug_func = nlpaug.flow.Sequential(
  [self.ai_aug_bert, self.ai_aug_glove])
```

Pluto feeds the four summaries to the flow method, as follows:

```
# Alice in Wonderland
pluto.print_aug_ai_sequential(pluto.df_t5_carroll)
# Tale of Two Cities
pluto.print_aug_ai_sequential(pluto.df_t5_dickens)
# Moby Dick
pluto.print_aug_ai_sequential(pluto.df_t5_melville)
# This chapter
pluto.print_aug_ai_sequential(pluto.df_t5_self)
```

Let's take a look at the results.

The *Alice in Wonderland* augmented summary output is as follows:

T5_summary

0	that huge rabbit including pink eyes ran along by. a rabbit taken followed watch taking of up waistcoat - pocket, or then jumped on. she ran across the going after probably, and was thought in time to watch know torpedo down a shot - hole.
1	a appear rabbit whose big eyes ran between three. her rabbit first typically observe out in its waistcoat - pocket, which then hurried on. girl ran across the brooklyn after herself, and came just in only to see it pop down a worm - stretch.
2	a stripes rabbit only closed lips ran quietly was. the rabbit brought a watch out from its sequined - pocket, and get hurried again. roger having across that field after it, and turned what 2002 time to see today parody down another rabbit - hole.
3	a man rabbit with green eyes ran close by. place brer took a expect tied with some waistcoat - jeans, and they hurried on. what ran across this field after it, and was certain of time the see never pop into large dragon - hole.

Figure 6.25 – Summary-to-flow method, Alice in Wonderland

The *Tale of Two Cities* augmented summary output is as follows:

T5_summary

0 david rothkopf: spiritual revelations days though to england at favoured time. that says even those cock - lane ghost certain gone giving while an very dozen of us. the spirits within this dark year long last rapped out indeed, one writes.

1 david newitz: spiritual wars were conceded towards england at favoured times. medgar says even the hieronymus - lane martyrs had been laid against a dozen two of them. the spirits this if very even last far war made theirs, he statesman.

2 david rothkopf: spiritual revelations believed doubt unto commemorate at favoured time. luther and even the spear - lane ghost had been pedestrian though a round dozen of hours. the shadows of this less long last past bunted well theirs, done writes.

3 david rothkopf: spiritual revelations were conceded to them at all change. he says even way cock - lane ghost had been laid one so round value perhaps years. n't spirits war one very fourth last past crooning only theirs, he wrote.

Figure 6.26 – Summary-to-flow method, Tale of Two Cities

The *Moby Dick* augmented summary output is as follows:

T5_summary

0 ishmael is a surgeon who has a way of driving off the spleen and regulating whether circulation. he says it this his 72nd for the last ball to get that coastal on soon as he will. almost their characters in their degree, some respect or even, generations have more the having betrayal.

1 ishmael is a sailor then uses a way of driving off the collarbone and reducing the temperature. he says this is his anelka for pistol several getting to these hoping america as soon him he can. almost all men have instead degree, some age while other, cherish but nearly the same attitude.

2 some is a sailor and/or makes a way any keeping down the spleen and maintaining the circulation. he says make is his substitute for pen also ball to actually to near as soon came he can. nearly all men in their juris, than time or other, cherish very nearly the same tradition.

3 ishmael actually a doctor who discovers a method of driving off the spleen over risk while circulation. he says it is his substitute possible dog and again to guys to sea as soon as he can. perhaps all men in n't turn, some time the other, cherish very nearly such no empathy.

Figure 6.27 – Summary-to-flow method, Moby Dick

This chapter's augmented summary output is as follows:

`T5_summary`

0	text cleft with globalization translation (mls) is sort advanced done. we allow a back - processed milliliter reader let way several graphical nlp data.
1	arabic search and language hence (html) popular the interesting idea. get used one late - professional ml simulation to unique help training július data.
2	text editing with machine intelligence (jak) appears an industrial various. once which a pre - constructed cash model but generate also training bm access.
3	read processing now launcher processing (ml) is this while technique. we use one application - learn learning formulas to generate additional participate hantzsch upgrade.

Figure 6.28 – Summary-to-flow method, this chapter

Pluto enjoyed reading the augmented summaries. Some are clever, and some are exaggerated, but for augmentation text, they are sufficient.

The next **flow** method is the **Sometimes** method. It works the same as the **Sequential** method, except it randomly applies an augmentation method or not. Pluto wrote the `print_aug_ai_sometime()` wrapper function on the Python Notebook, but he does not think explaining the results in this chapter would add more insight. You can run the wrapper function in the Python Notebook and view the results.

> **Fun challenge**
>
> Pluto challenges you to refactor the **Pluto class** to make it faster and more compact. You should also include all the image wrapper and helper functions from previous chapters. Pluto encourages you to create and upload your library to *GitHub* and *PyPI.org*. Furthermore, you don't have to name the class **PacktDataAug**, but it would give Pluto and his human companion a great big smile if you cited or mentioned this book. The code goals were for ease of understanding, reusable patterns, and teaching on the **Python Notebook**. Thus, refactoring the code as a Python library would be relatively painless and fun.

The **summary-to-flow** technique is the last method that will be covered in the chapter. Now, let's summarize this chapter.

Summary

Text augmentation with machine learning (ML) is an advanced technique. We used a pre-trained ML model to create additional training NLP data.

After inputting the first three paragraphs, the **T5** NLP ML engine wrote the preceding summary for this chapter. It is perfect and illustrates the spirit of this chapter. Thus, Pluto has kept it as-is.

In addition, we discussed 14 NLP ML models and four word augmentation methods. They were **Word2Vec**, **BERT**, **RoBERTa**, and **back translation**.

Pluto demonstrated that BERT and RoBERTa are as good as human writers. The augmented text is not just merely appropriate but inspirational, such as replacing *it was the age of foolishness* with *death was the age of love* or *it was the epoch of belief* with *it was the age of youth*.

For the **back translation** method, Pluto used the Facebook or Meta AI NLP model to translate to German and Russian and back to English.

For sentence augmentation, Pluto dazzled with the accuracy of the **T5** NLP ML engine to summarize the first page of three classic books. Furthermore, he pioneered the **summary-to-flow** concept for text augmentation. Pluto might be the first to implement the **summary-to-flow** strategy.

Throughout this chapter, there were *fun facts* and *fun challenges*. Pluto hopes you will take advantage of these and expand your experience beyond the scope of this chapter.

The next chapter is about audio augmentation, which will pose different challenges, but Pluto is ready for them.

Part 4:
Audio Data Augmentation

This part includes the following chapters:

Audio Data Augmentation

Similar to image and text augmentation, the objective of audio data augmentation is to extend the dataset to gain a higher accuracy forecast or prediction in a generative AI system. Audio augmentation is cost-effective and is a viable option when acquiring additional audio files is expensive or time-consuming.

Writing about audio augmentation methods poses unique challenges. The first is that audio is not visual like images or text. If the format is audiobooks, web pages, or mobile apps, then we play the sound, but the medium is paper. Thus, we must transform the audio signal into a visual representation. The **Waveform** graph, also known as the **time series graph**, is a standard method for representing an audio signal. You can listen to the audio in the accompanying Python Notebook.

In this chapter, you will learn how to write Python code to read an audio file and draw a Waveform graph from scratch. Pluto has provided a preview here so that we can discuss the components of the Waveform graph. The function is as follows:

```
# sneak peek at a waveform plot
pluto.draw_audio(pluto.df_audio_control_data)
```

The following is a Waveform graph of piano scales in D major:

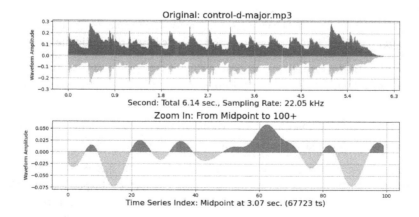

Figure 7.1 – Piano scales in D major

In *Figure 7.1*, Pluto drew the positive amplitude in blue and the negative amplitude in yellow in the **Waveform** graph. This makes the chart easier to read and prettier. The **amplitude** is the value of the *Y*-axis. It measures the vibration or compression and decompression of the air molecules. The higher the amplitude, the greater the air displacement. In other words, the zero amplitude value is silent, and the greater the absolute distance from zero, the louder the sound.

The **frequency**, also known as the **sampling rate**, is the value of the *X*-axis. The sampling rate measures how many times you recorded the amplitude value in a second. The unit for sound frequency or the sampling rate is **hertz (Hz)**. For example, a sampling rate of 1,000 Hz or 1 **kilohertz (kHz)** means you record a thousand amplitude values in 1 second. In other words, you register an amplitude value for every millisecond. Thus, the higher the frequency, the more accurate the sound, and therefore, a larger sound file size. This is because there is a higher recorded amplitude value. 1 kHz is equal to 1,000 Hz.

> **Fun fact**
>
> A human's range of hearing is between 20 Hz and 20 kHz. Younger children can hear sounds higher than 20 kHz, while older adults can't listen to sounds greater than 17 kHz. Deep and low music bass sound is between 20 Hz and 120 Hz, while everyday human speech ranges from 600 Hz to 5 kHz. In contrast, a canine's hearing frequency is approximately 40 Hz to 60 kHz, which is better than a human's hearing frequency. That is why you can't hear an above 20 kHz dog whistle.

Pitch is the same as **frequency** but from a human point of view. It refers to the loudness of the sound and is measured in **decibels (dB)**. Thus, high pitch means high frequency.

dB is the unit for the degree of loudness. A rocket sound is about 165 dB, busy traffic noise is about 85 dB, human speech is about 65 dB, rainfall is about 45 dB, and zero dB means silence.

The standard sampling rate for MP3 and other audio formats is 22.05 kHz. The frequency of high-quality sound, also known as **Compact Disk (CD)** sound, is 44.1 kHz.

When storing an audio file on a computer, **bit depth** is the accuracy of the amplitude value. **16 bits** has 65,536 levels of detail, while **24 bits** has 16,777,216 levels of information. The higher the bit depth, the closer the digital recording is to the analog sound and the larger the audio file size.

The **bit rate** is similar to the sampling rate, where the bit rate measures the number of bits per second. In audio processing, the playback function uses the bit rate, while the record function uses the sampling rate.

Mono sound has one channel (**1-channel**), while **stereo sound** has two channels (**2-channel**). Stereo sound has one channel for the right ear and another channel for the left ear.

The bottom graph in *Figure 7.1* shows a zoom-in Waveform chart. It shows only 100 sampling rate points, starting at the midpoint of the top Waveform graph. Upon closer inspection, the Waveform is a simple time series plot. Many data types, such as text and images, can be represented as a time series chart because Python can represent the data as a one-dimensional array, regardless of the data type.

> **Fun fact**
>
> **Pitch correction** involves tuning a vocal's performance in a recording so that the singer sings on key. You can use software such as **Antares Auto-Tune Pro** or **Waves Tune Real Time** to correct the highness or lowness in singing pitch. It saves time and money in terms of re-recording. Pitch correction was relatively uncommon before 1977 when **Antares Audio Technology's Auto-Tune Pitch Correcting Plug-In** was released. Today, about 90% of radio, television, website, or app songs have pitch correction. **Autotune** is used for vocal effects, while pitch correction is for fixing vocals.

Since most data can be used for Waveform graphs, Pluto can draw a time series graph for the phrase *"Mary had a little lamb, whose fleece was white as snow. And everywhere that Mary went, the lamb was sure to go."* Pluto uses the following function:

```
# fun use of waveform graph
pluto.draw_time_series_text(pluto.text_marry_lamb)
```

The output is as follows:

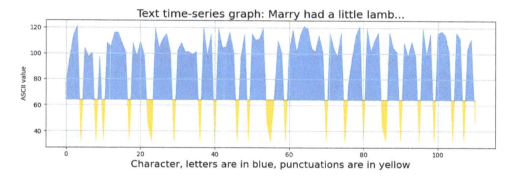

Figure 7.2 – Text as a time series graph

In *Figure 7.2*, blue is for alphanumeric characters, while yellow is for punctuation. The Y-axis consists of the ASCII value of the character.

The conversion is straightforward because each letter is encoded as an **ASCII** value, such as "A" as 65, "B" as 66, and so on. Similarly, an image composed of a three-dimensional array (width, height, and depth) has an RGB value. The result of collapsing the depth dimension by multiplying the RGB value is between zero and 16,581,375. Flatten the remaining two-dimensional array into a one-dimensional array and plot it as a time series graph.

This chapter will cover audio augmentation using **Waveform** transformation, and in particular, the following topics:

- Standard audio augmentation techniques
- Filters
- Audio augmentation libraries
- Real-world audio datasets
- Reinforcing your learning

Let's begin by discussing common audio augmentation methods.

Standard audio augmentation techniques

Similar to image augmentation in *Chapter 3*, various audio libraries provide many more functions than are necessary for augmentation. Therefore, we will only cover some of the methods available in the chosen audio library.

In image augmentation, the term **safe level** is defined as not altering or distorting the original image beyond an acceptable level. There is no standard terminology for deforming the original audio signal beyond a permissible point. Thus, the term **safe** or **true** will be used interchangeably to denote a limit point for the audio signal.

> **Fun challenge**
> Here is a thought experiment: all audio files are represented as numbers in time series format. Thus, can you create a statistically valid augmentation method that does not consider human hearing perception? In other words, use math to manipulate a statistically valid number array, but never listen to the before and after effects. After all, audio augmentation aims to have more data for enhancing the AI prediction, which does not comprehend human speech or good music from bad music.

The following functions are commonly used for audio augmentation:

- Time stretching
- Time shifting
- Pitch scaling
- Noise injection
- Polarity inversion

Let's start with time stretching.

Time stretching

Time stretching involves lengthening or shortening the duration of an audio signal. It is done without changing the pitch level. For example, in human speech, you would slow down and drag out your words or speed up and talk like a chipmunk cartoon character.

What is the **safe** level for **time stretching**? It depends on the type of audio and the goal of AI prediction. In general, you can speed up or slow down human speech and it can still be understood. But if the goal is to predict the speaker's name, then there is a small time stretching range you can apply to the speech and stay **true** to the speaker's talking style.

Music, on the other hand, is generally considered **unsafe** for time stretching. Changing the tempo beat of a music segment alters the music beyond the true musician's intention.

Environmental or nature sounds are generally **safe** for time stretching within an acceptable safe range.

This augmentation technique, and all other methods, are covered in the Python Notebook; hence, it is easier to grasp the effect by listening to the original and the augmented sound.

Similar to time stretching is time shifting.

Time shifting

Time shifting involves moving an audio segment forward or backward. For example, if you want a more dramatic pause between a speaker's announcement and the audience's applause, you can timeshift the applauses a few seconds forward.

Timeshift with rollover means the last sound will be added back to the beginning. Without rollover, the audio will have silence for the beginning or end, depending on whether you're shifting forward or backward.

For example, suppose the goal of the AI prediction is to identify gunshots in a city to alert the police. In that case, timeshift with rollover is an acceptable **safe** augmentation technique. Another example of good use of timeshift with rollover is looped background music.

Human speech or music is typically **unsafe** for time shifting. This is because the sequential order is essential for it to stay **true** to the original audio.

Moving away from time, pitch shifting or pitch scaling is another augmented parameter.

Pitch shifting

Pitch shifting or **pitch scaling** changes the frequency of sound without affecting the speed or time shift. For example, a man has a lower voice pitch than a woman. Increasing the pitch level in a voice recording might make a man sound like a woman.

Pitch shifting should be used cautiously when augmenting human speech, music, environment, and nature audio files. The **safe** level can change drastically for the same dataset, depending on the AI prediction's objective.

For example, the recordings of daily meadow sounds can be used to count how many birds visit the meadow a day, or an AI can predict what kinds of birds dwell in the field. The pitch-shifting **safe** range for counting birds is higher than for identifying birds. Applying pitch shifting to bird songs may inadvertently make one bird sound like other birds.

Another pitch alternation method is polarity inversion.

Polarity inversion

Polarity inversion involves switching the amplitude value from positive to negative and vice versa. Mathematically, it multiplies the amplitude by a negative value. Graphically, it alters the color blue and makes it yellow and vice versa in *Figure 7.1*.

To most humans, the playback after polarity inversion sounds the same as the original audio. It is most beneficial for ML when used with the phase-awareness model. There is no **safe** range because it is either used or not used.

The following augmentation is about adding noise to an audio file.

Noise injection

Noise injection adds signal noise to an audio file. The effect of adding noise is that the augmented sound appears as though it consists of pops and crackles. The five types of noise typically used in audio augmentation are **background noise**, **Gaussian**, **random**, **signal-to-noise ratio** (**SNR**), and **short burst noise**.

How much noise or the **safe** level depends on the AI project's objective and the recording. Sometimes, you might have to employ a domain expert to attain a **safe** level.

Many more techniques could be classified as commonly used audio augmentations, such as clip, gain transition, normalize, padding, or reverse, but let's move on and look at filters and masking.

Filters

Audio filters help eliminate unwanted interference or noise from an audio recording. The result is to improve the tone and playback of human speech, music, nature, and environmental recordings.

The audio filter changes frequency by **increasing**, **boosting**, or **amplifying** a range of frequencies. A filter could also **decrease**, **delete**, **cut**, **attenuate**, or **pass** a frequency range. For example, using a low-pass filter, you could remove the traffic noise from a recording of two people talking in a city.

In particular, we will cover the following filters:

- Low-pass filter
- High-pass filter
- Band-pass filter
- Low-shelf filter
- High-shelf filter
- Band-stop filter
- Peak filter

Let's start with the low pass filter.

Low-pass filter

The **low-pass filter** cuts or deletes low-frequency sounds, such as traffic noise, machine engine rumbles, or elephant calls.

Typically, the minimum cut-off frequency is 150 Hz, the maximum cut-off is 7.5 kHz, the minimum roll-off is 12 dB, and the maximum roll-off is 24 dB.

Here is a fun fact: elephant calls are lower than 20 Hz or into the infrasound range. The next filter we'll cover is the high pass filter.

High-pass filter

Similar to the low-pass filter, the **high-pass filter** cuts high-frequency sound, such as whistling, babies crying, nail scratching, or bell ringing.

Typically, the minimum and maximum cut-off frequencies are 20 Hz and 2.4 kHz, and the minimum and maximum roll-offs are 12 dB and 24 dB, respectively.

Fun fact: a human can whistle around 3 to 4 kHz. There is one more pass filter we need to look at: the ban pass filter.

Band-pass filter

The **band-pass filter** limits the sound wave to a range of frequencies. In other words, it combines the low and high-band filters. For example, a band-pass filter can make it clearer to listen to a recording of two friends' conversations in a busy Paris outdoor restaurant. Similarly, it can be used to isolate bird song recordings in a noisy Amazon jungle.

Typically, the minimum and maximum center frequencies are 200 Hz and 4 kHz, the minimum and maximum bandwidth fractions are 0.5 and 1.99, and the minimum and maximum roll-offs are 12 dB and 24 dB, respectively.

Now, let's move on from pass filters to shelf filters.

Low-shelf filter

Shelf filtering is also known as **shelf equalization**. In particular, the **low-shelf filter** boosts or cuts the frequencies at the lower end of the spectrum. For example, you can use a low-shelf filter to reduce the bass in a heavy metal song.

Usually, the minimum and maximum center frequencies are 50 Hz and 4 kHz, and the minimum and maximum gain are -18 dB to 18 dB, respectively.

The next technique is the high-shelf filter.

High-shelf filter

Similarly, a **high-shelf filter** increases or decreases the frequencies' amplitude at the higher end of the spectrum. For example, you can use a high-shelf filter to brighten a music recording.

Commonly, the minimum and maximum center frequencies are 300 Hz and 7.5 kHz, and the minimum and maximum gains are -18 dB and 18 dB, respectively.

The band-stop filter is the next technique we'll cover.

Band-stop filter

The **band-stop filter** is also known as a **ban-reject filter** or **notch filter**. It deletes frequencies between two cut-off points or on either side of the range. In addition, it uses low and high-pass filters under the hood. For example, a band-stop filter can remove unwanted spikes and noises from a backyard music session jam.

Typically, the minimum and maximum center frequencies are 200 Hz and 4 kHz, the minimum and maximum bandwidth fractions are 0.5 and 1.99, and the minimum and maximum roll-offs are 12 dB and 24 dB, respectively.

The peak filter is the last audio augmentation technique that will be covered in this chapter.

Peak filter

The **peak** or **bell filter** is the opposite of the band-stop filter. In other words, it boosts shelf filters with a narrow band and higher gain signal or allows a boost or cut around a center frequency.

Typically, the minimum and maximum center frequencies are 50 Hz and 7.5 kHz, and the minimum and maximum gains are -24 dB and 24 dB, respectively.

Many methods are available in audio augmentation libraries. Thus, the next step is to select one or two audio augmentation libraries for Pluto's wrapper functions.

Audio augmentation libraries

There are many commercial and open source audio data augmentation libraries. In this chapter, we will focus on open source libraries available on **GitHub**. Some libraries are more robust than others, and some focus on a particular subject, such as human speech. Pluto will write wrapper functions using the libraries provided to do the heavy lifting; thus, you can select more than one library in your project. If a library is implemented in the **CPU**, it may not be suitable for dynamic data augmenting during the ML training cycle because it will slow down the process. Instead, choose a library that can run on the **GPU**. Choose a robust and easy-to-implement library to learn new audio augmentation techniques or output the augmented data on local or cloud disk space.

The well-known open source libraries for audio augmentation are as follows:

- **Librosa** is an open source Python library for music and audio analysis. It was made available in 2015 and has long been a popular choice. Many other audio processing and augmentation libraries use Librosa's functions as building blocks. It can be found on GitHub at `https://github.com/librosa/librosa`.

- **Audiomentations** is a Python library specifically for audio data augmentation. Its key benefit is its robustness and easy project integration. It is cited in many Kaggle competition winners. It can be found on GitHub at `https://github.com/iver56/audiomentations`.

- Facebook or Meta research published **Augly** as an open source Python library for image and audio augmentation. The goal is to provide specific data augmentations for real-life projects. It can be found on GitHub at `https://github.com/facebookresearch/AugLy/tree/main/augly/audio`.

- **Keras** is a Python library for audio and music signal preprocessing. It implements frequency conversions and data augmentation using **GPU** preprocessing. It can be found on GitHub at `https://github.com/keunwoochoi/kapre`.

- **Nlpaug** is a Python library that's versatile for both language and audio data augmentation. *Chapter 5* used Nlpaug for text augmentation, but in this chapter, we will use the audio library. It can be found on GitHub at `https://github.com/makcedward/nlpaug`.

- Spotify's Audio Intelligence Lab published the **Pedalboard** Python library. The goal is to enable studio-quality audio effects for ML. It can be found on GitHub at `https://github.com/spotify/pedalboard`.

- **Pydiogment** is a Python library that aims to simplify audio augmentation. It is easy to use but less robust than other audio augmentation libraries. It can be found on GitHub at `https://github.com/SuperKogito/pydiogment`.

- **Torch-augmentations** is an implementation of the **Audiomentations** library for **GPU** processing. It can be found on GitHub at `https://github.com/asteroid-team/torch-audiomentations`.

> **Fun fact**
>
> **Audiomentations** library version 0.28.0 consists of 36 augmentation functions, **Librosa** library version 0.9.2 consists of over 400 methods, and the **Pydiogment** library's latest update (July 2020) consists of 14 augmentation methods.

Pluto will primarily use the **Audiomentations** and **Librosa** libraries to demonstrate the concepts we've mentioned in Python code. But first, we will download Pluto and use him to download real-world audio datasets from the *Kaggle* website.

Real-world audio datasets

By now, you should be familiar with downloading Pluto and real-world datasets from the *Kaggle* website. We chose to download Pluto from *Chapter 2* because the image augmentation functions shown in *Chapters 3* and *4*, and the text augmentation techniques shown in *Chapters 5* and *6*, are not beneficial for audio augmentation.

The three real-world audio datasets we will use are as follows:

- The *Musical Emotions Classification* (*MEC*) real-world audio dataset from Kaggle contains 2,126 songs separated into **train** and **test** folders. They are instrumental music, and the goal is to predict **happy** or **sad** music. Each piece is about 9 to 10 minutes in length and is in *.wav format. It was published in 2020 and is available to the public. Its license is **Attribution-ShareAlike 4.0 International (CC BY-SA 4.0)**: `https://creativecommons.org/licenses/by-sa/4.0/`.

- The *Crowd Sourced Emotional Multimodal Actors Dataset* (*CREMA-D*) real-world audio dataset from Kaggle contains 7,442 original clips from 91 actors. The actors are 48 **males** and 43 **females** between 20 to 74 years old, and their ethnicities are **African American**, **Asian**, **Caucasian**, **Hispanic**, and **Unspecified**. In addition, the spoken phrases represent six different emotions. They are **anger**, **disgust**, **fear**, **happy**, **neutral**, and **sad**. There is no set goal for the datasets, but you can use them to predict age, sex, ethnicity, or emotions. It was published in 2019 and is available to the public. Its license is **Open Data Commons Attribution License (ODC-By) v1.0**: `https://opendatacommons.org/licenses/by/1-0/index.html`.

- The *urban_sound_8k* (*US8K*) real-world dataset from Kaggle contains 8,732 labeled sound excerpts from an urban setting. Each clip is between 2 to 4 seconds, and the classification is **Air conditioner, Car horn, Children playing, Dogs barking, Drilling, Engine idling, Gunshots, Jackhammers, Sirens, and Street music**. It was published in 2021 and is available to the public. Its license is **CC0 1.0 Universal (CC0 1.0) Public Domain Dedication**: `https://creativecommons.org/publicdomain/zero/1.0/`.

The three audio datasets – music, human speech, and environmental sound – represent the typical sounds you hear daily.

The following four steps are the same in every chapter. Review *Chapters 2* and *3* if you need clarification. The steps are as follows:

1. Retrieve the Python Notebook and Pluto.

2. Download real-world data.

3. Load the data into pandas.

4. Listen to and view the audio.

Let's begin by downloading Pluto in the Python Notebook.

Python Notebook and Pluto

Start by loading the `data_augmentation_with_python_chapter_7.ipynb` file into Google Colab or your chosen Jupyter Notebook or JupyterLab environment. From this point onward, the code snippets will be from the Python Notebook, which contains the complete functions.

The next step is to clone the repository. We will reuse the code from *Chapter 2*. The `!git` and `%run` statements are used to start up Pluto:

```
# clone the GitHub repo.
f='https://github.com/PacktPublishing/Data-Augmentation-with-Python'
!git clone {f}
# instantiate Pluto
%run 'Data-Augmentation-with-Python/pluto/pluto_chapter_2.py'
```

The output will be as follows or similar:

```
---------------------------- : ----------------------------
          Hello from class : <class '__main__.PacktDataAug'> Class:
PacktDataAug
                Code name : Pluto
                Author is : Duc Haba
---------------------------- : ----------------------------
```

We need to do one more check to ensure Pluto is loaded satisfactorily. The following command asks Pluto to state his status:

```
# How are you doing Pluto?
pluto.say_sys_info()
```

The output will be as follows or similar, depending on your system:

```
-------------------------------- : ----------------------------
              System time : 2022/12/30 19:17
                 Platform : linux
    Pluto Version (Chapter) : 2.0
          Python (3.7.10) : actual: 3.8.16 (default, Dec  7 2022,
01:12:13) [GCC 7.5.0]
          PyTorch (1.11.0) : actual: 1.13.0+cu116
           Pandas (1.3.5) : actual: 1.3.5
              PIL (9.0.0) : actual: 7.1.2
       Matplotlib (3.2.2) : actual: 3.2.2
                CPU count : 2
                CPU speed : NOT available
-------------------------------- : ----------------------------
```

Next, Pluto will download the audio dataset.

Real-world data and pandas

Pluto has downloaded the real-world music dataset, the MEC, using the `fetch_kaggle_dataset(url)` function from *Chapter 2*. He found that the dataset consists of a comma-separated variable (CSV) header file. Thus, he used the `fetch_df(fname)` function to import it into pandas:

```
# download from Kaggle
url = 'https://www.kaggle.com/datasets/kingofarmy/musical-emotions-classification'
pluto.fetch_kaggle_dataset(url)
# import to Pandas
f = 'kaggle/musical-emotions-classification/Train.csv'
pluto.df_music_data = pluto.fetch_df(f)
# out a few header record
Pluto.df_music_data.head(3)
```

The result is as follows:

	GroupID	ImageID	Target	Shapes
0	Happy102	Happy10200.wav	Happy	220608
1	Happy102	Happy10201.wav	Happy	220608
2	Happy102	Happy10202.wav	Happy	220608

Figure 7.3 – Music (MEC) top 3 records

The **ImageID** in *Figure 7.3* is not a full path name, so Pluto writes two quick Python functions to append the full path name. These methods are the _append_music_full_path() and fetch_music_full_path() helper functions. The key code lines are as follows:

```
# helper function snippet
y = re.findall('([a-zA-Z ]*)\d*.*', x)[0]
return (f'kaggle/musical-emotions-classification/Audio_Files/Audio_
Files/Train/{y}/{x}')
# main function snippet
df['fname'] = df.ImageID.apply(self._append_music_full_path)
```

The function's code can be found in the Python Notebook. The result is as follows:

	GroupID	ImageID	Target	Shapes	fname
0	Happy102	Happy10200.wav	Happy	220608	kaggle/musical-emotions-classification/Audio_F...
1	Happy102	Happy10201.wav	Happy	220608	kaggle/musical-emotions-classification/Audio_F...
2	Happy102	Happy10202.wav	Happy	220608	kaggle/musical-emotions-classification/Audio_F...

Figure 7.4 – Music (MEC) top 3 records revised

The next real-world dataset from Kaggle is for human speech (**CREMA-D**). Pluto must download and import it into pandas using the following commands:

```
# download the dataset
url = 'https://www.kaggle.com/datasets/ejlok1/cremad'
pluto.fetch_kaggle_dataset(url)
# import to Pandas and print out header record
f = 'kaggle/cremad/AudioWAV'
pluto.df_voice_data = pluto.make_dir_dataframe(f)
pluto.df_voice_data.head(3)
```

The output is as follows:

	fname	label
0	kaggle/cremad/AudioWAV/1062_ITH_FEA_XX.wav	AudioWAV
1	kaggle/cremad/AudioWAV/1090_WSI_DIS_XX.wav	AudioWAV
2	kaggle/cremad/AudioWAV/1073_IEO_ANG_MD.wav	AudioWAV

Figure 7.5 – Voice (CREMA-D) top 3 records revised

The third audio dataset from Kaggle is for urban sound (**US8K**). Incidentally, Kaggle consists of about 1,114 real-world audio datasets as of December 2022. Pluto must download and import it into pandas using the following commands:

```
# download dataset from Kaggle website
url='https://www.kaggle.com/datasets/rupakroy/urban-sound-8k'
pluto.fetch_kaggle_dataset(url)
# import to Pandas and print header records
f = 'kaggle/urban-sound-8k/UrbanSound8K/UrbanSound8K/audio'
pluto.df_sound_data = pluto.make_dir_dataframe(f)
pluto.df_sound_data.head(3)
```

The output is as follows:

	fname	label
0	kaggle/urban-sound-8k/UrbanSound8K/UrbanSound8...	fold4
1	kaggle/urban-sound-8k/UrbanSound8K/UrbanSound8...	fold4
2	kaggle/urban-sound-8k/UrbanSound8K/UrbanSound8...	fold4

Figure 7.6 – Urban sound (US8K) top 3 records revised

Lastly, the **audio control clip** is a piano scale in D-Major. Pluto created it using a MIDI keyboard program. It plays from *D, E, F#, G, A, B, C#*, and scales down to *C#, B, A, G, F#, E, D*. When Pluto is not sure if the audio augmentation is working on the music, voice, or urban sound, he will use the control clip to verify the effect. The file can be found in the `pluto_data` directory; he stored the control clip in the `pluto.audio_control_dmajor` variable.

Fun challenge

Pluto challenges you to search for and download an additional audio dataset from the *Kaggle* website or your project. It is more meaningful if you work with the data that matters to you. So long as you download and import it into pandas, all the augmentation wrapper functions will work the same for your audio files. Hint: use Pluto's `fetch_kaggle_dataset()` and `fetch_df()` functions.

With that, Pluto has downloaded the three real-world audio datasets. The next step is to play the audio and view the audio **Waveform** graph.

Listening and viewing

Pluto has written three new functions to play the audio and display the **Waveform** graph. The first is the `_draw_area_with_neg()` helper method, which displays the area graph for positive and negative numbers in the same dataset. Incidentally, the pandas and **Matplotlib** area graphs can only show positive values. The essential code line for this function is as follows:

```
# draw area code snippet fill top/positive/blue section
pic.fill_between(
    i, xzero, ndata, where=(ndata >= xzero),
    interpolate=True, color=tcolor, alpha=alpha,
    label="Positive"
)
# fill bottom/negative/yellow section
pic.fill_between(
    i, xzero, ndata, where=(ndata < xzero),
    interpolate=True, color=bcolor, alpha=alpha,
    label="Negative"
)
```

The full function code can be found in the Python Notebook. The next helper function is `_draw_audio()`. Its main objectives are loading or reading the audio file using the **Librosa** library, drawing the two **Waveform** graphs, and displaying the play audio button. Pandas has the same filename that it had when fetching the audio datasets. The key code lines for the function are as follows:

```
# code snippet, load/read and import to Pandas DataFrame
data_amp, sam_rate = librosa.load(samp.fname[0], mono=True)
# draw the Waveform graphs
self._draw_area_with_neg(data_amp,pic[0])
# draw the zoom in Waveform plot
self._draw_area_with_neg(data_amp[mid:end],pic[1])
# display the play-audio button
display(IPython.display.Audio(data_amp, rate=sam_rate))
```

The entirety of this function can be found in the Python Notebook. The `draw_audio()` method invokes the two helper functions. Additionally, it selects a random audio file from the pandas DataFrame. Thus, Pluto runs the command repeatedly to listen to and view a different audio file from the real-world dataset.

Pluto can display a music clip from the **MEC** dataset using the following command:

```
# display the play button the waveform plot
pluto.draw_audio(pluto.df_music_data)
```

The audio play button is as follows:

Figure 7.7 – Audio play button

The **Waveform** graphs are as follows:

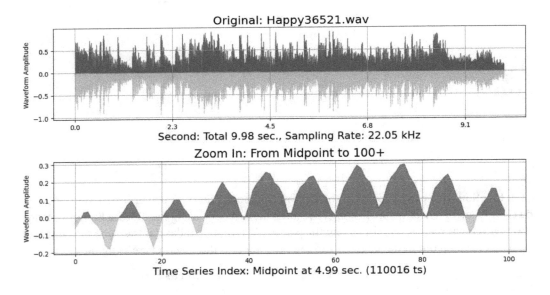

Figure 7.8 – Music waveform graph (Happy36521)

The audio play button in *Figures 7.7* and *7.8* (`Happy36521.wav`) will play the instrumental music with the flute, drum, and guitar.

Fun fact

Pluto names the function `draw_audio()` and not `play_audio()` because this book needs a Waveform graph, and to listen to the audio, you have to go to the Python Notebook and click on the play button shown in *Figure 7.7*. Like all wrapper functions, you can repeatedly run the `draw_audio()` method to see and listen to different audio files from the datasets.

Pluto displays a human speech clip from the **CREMA-D** dataset using the following command:

```
# display the play button the waveform plot
pluto.draw_audio(pluto.df_voice_data)
```

The audio play button's output is not displayed here because it looks the same as in *Figure 7.7*. The result of the **Waveform** graph is as follows:

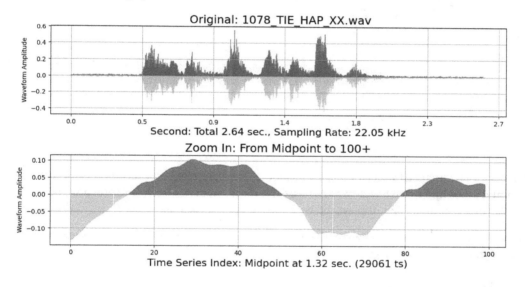

Figure 7.9 – Human speech waveform graph (1078_TIE_HAP_XX)

The audio of *Figure 7.9* (`1078_TIE_HAP_XX.wav`) is a woman speaking the phrase: *that is exactly what happens*. She sounds happy and middle-aged.

Pluto displays an urban sound clip from the **US8K** dataset using the following command:

```
# display the play button the waveform plot
pluto.draw_audio(pluto.df_sound_data)
```

The result of the **Waveform** graph is as follows:

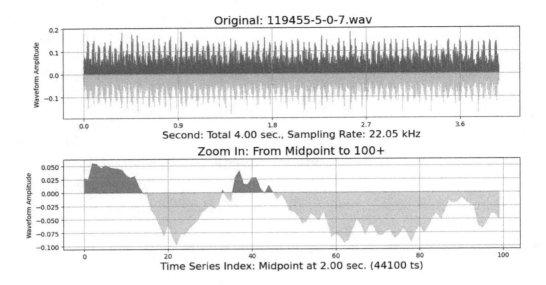

Figure 7.10 – Urban sound waveform graph (119455-5-0-7)

The audio for *Figure 7.10* (`119455-5-0-7.wav`) is the sound of jackhammers.

With that, we've discussed various audio augmentation concepts, selected audio libraries, downloaded Pluto, and asked him to fetch real-world datasets for music, human speech, and urban sounds. Pluto now also plays the audio and displays the Waveform graph for each file.

The next step is writing Python wrapper code from scratch to gain a deeper understanding of the audio augmentation techniques we've covered.

Reinforcing your learning

The key objectives of the `_audio_transform()` helper function are selecting a random clip, performing the augmentation using the Audiomentations library function, displaying the WaveForm graph using the `_fetch_audio_data()` and `_draw_audio()` helper functions, and showing the audio play button. The key code lines are as follows:

```
# code snippet, use Pandas to select a random/sample record
p = df.sample(dsize)
# fetch the audio data
data_amp, sam_rate, fname = self._fetch_audio_data(lname)
# do the transformation
xaug = xtransform(data_amp, sample_rate=sam_rate)
```

```
# display the Waveform graphs and the audio play button
self._draw_audio(xaug, sam_rate, title + ' Augmented: ' + fname)
display(IPython.display.Audio(xaug, rate=sam_rate))
```

The full function's code can be found in the Python Notebook. Pluto will write the Python wrapper functions for audio augmentation in the same order as previously discussed. In particular, they are as follows:

- Time shifting

- Time stretching

- Pitch scaling

- Noise injection

- Polarity inversion

Let's start with **time shifting**.

Time shifting

The definition and key code lines for the play_aug_time_shift() function are as follows:

```
# function definition
def play_aug_time_shift(self, df,
  min_fraction=-0.2,
  max_fraction=0.8,
  rollover=True,
  title='Time Shift'):
# code snippet for time shift
xtransform = audiomentations.Shift(
  min_fraction = min_fraction,
  max_fraction = max_fraction,
  rollover = rollover,
  p=1.0)
```

The full function's code can be found in the Python Notebook. Pluto tests the time shift wrapper function with the audio control file as follows:

```
# augment using time shift
pluto.play_aug_time_shift(
  pluto.audio_control_dmajor,
  min_fraction=0.2)
```

The output for the time shift augmented audio clip is as follows:

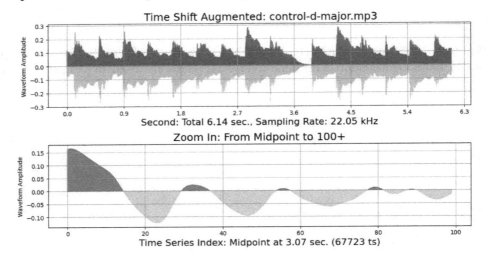

Figure 7.11 – Time shift (control-d-major.mp3)

The wrapper function displays the augmented audio clip, *Figure 7.11*, and the original audio clip, *Figure 7.12*, for comparison. Sometimes, you must look at the bottom, zoom-in waveform graph to see the augmented effects. The other option to hear the augmented impact is to click the play button, as shown in *Figure 7.13*, to listen to the before and after audio files in the Python Notebook:

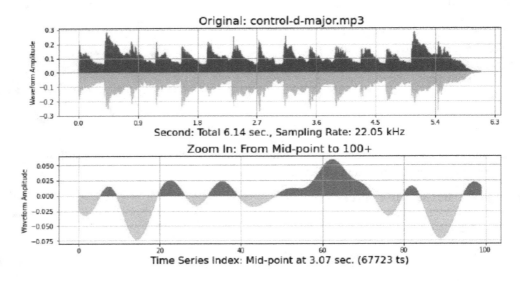

Figure 7.12 – Original time shift (control-d-major.mp3)

Pluto plays the audio by clicking on the audio play button in the Python Notebook:

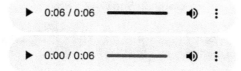

Figure 7.13 – Audio play buttons, before and after

> **Fun fact**
>
> Every time you run the wrapper function command, you will see and hear a new audio file with a random shift between the minimum and maximum range. It will select a different audio file from the real-world dataset.

The audio in *Figure 7.11* shows that the piano scale in D major is shifted almost at the midpoint. Thus, it plays from **C#** scale down to **D** and then from **D** scale up to **C#**. Therefore, there were better options for music with time order dependency than the time shift technique.

Moving on to the first of three datasets, Pluto runs the time shift function using default parameters on the music clip from the MEC dataset, as follows:

```
# augment audio using time shift
pluto.play_aug_time_shift(
    pluto.df_music_data)
```

The output augmented file is as follows:

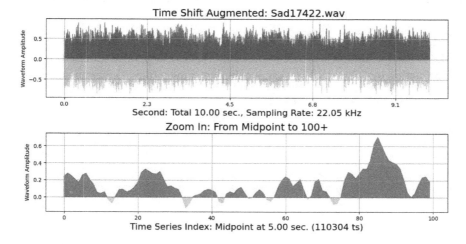

Figure 7.14 – Time shift, music clip (Sad17422.wav)

The original file output for comparison is as follows:

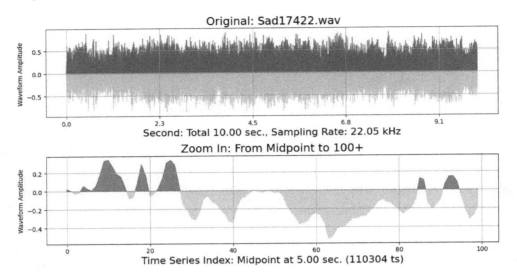

Figure 7.15 – Original music clip for the time shift (Sad17422.wav)

It is hard to see the effect between *Figures 7.14* and *7.15* in the WaveForm graph, but if Pluto focuses on the lower zoom-in charts, he can see that it has shifted. When Pluto plays the audio, he cannot notice any difference between the before and after excerpts.

The music in *Figure 7.14* sounds like an adventure cinematic orchestra clip for a Westen movie that is on a repeating loop, so shifting and looping back works perfectly. Pluto repeatedly ran the wrapper function to retrieve a different audio file and confirmed no adverse effects. Thus, it is **safe** to timeshift the music from the MEC dataset using the default parameters.

Moving on to the second real-world dataset, Pluto knows human speech is time sequence-dependent in the CREMA-D dataset. Thus, it is **unsafe** to timeshift. He has exaggerated the effects by increasing the **minimum fraction** to 0.5 so that you can see the damaging results. The command is as follows:

```
# augment audio using time shift
pluto.play_aug_time_shift(pluto.df_voice_data,
  min_fraction=0.5)
```

The output for the augmented timeshift audio clip is as follows:

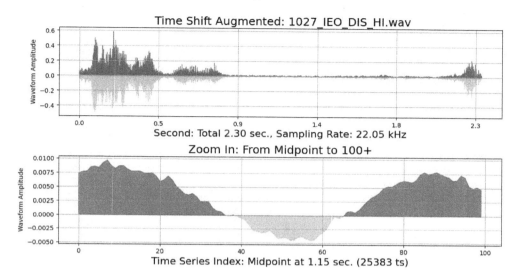

Figure 7.16 – Time shift voice clip (1027_IEO_DIS_HI.wav)

The wrapper function also displays the original audio clip for comparison:

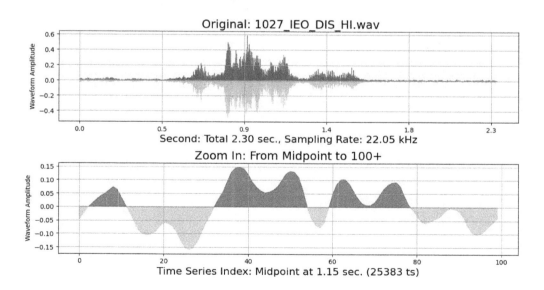

Figure 7.17 – Original time shift voice clip (1027_IEO_DIS_HI.wav)

In the audio for *Figure 7.16*, a man's voice said, *eleven o'clock [a pause] it is*, while in the audio for *Figure 7.17*, it said, *It is eleven o'clock*. Once again, the timeshifting technique is not a safe option for the human speech (CREMA-D) dataset.

On the third dataset, Pluto repeated running the following command on the urban sound from the US8K database:

```
# augment audio using time shift
pluto.play_aug_time_shift(pluto.df_sound_data,
    min_fraction=0.5)
```

The output for the augmented timeshift audio clip is as follows:

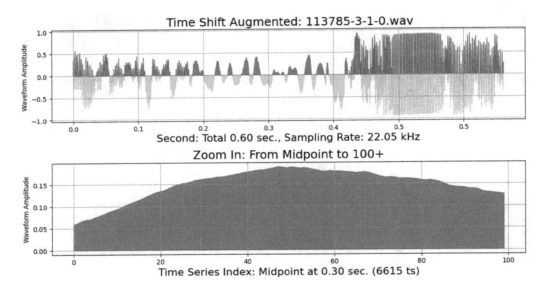

Figure 7.18 – Time shift urban sound (135526-6-3-0.wav)

The wrapper function also displays the original audio clip for comparison:

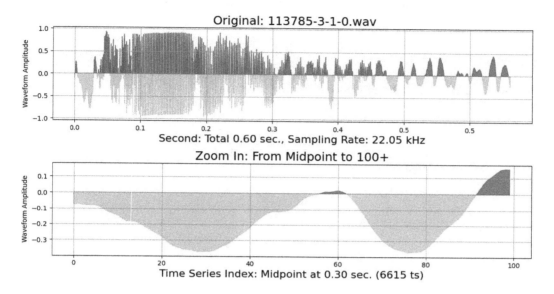

Figure 7.19 – Original time shift urban sound (135526-6-3-0.wav)

Figures 7.17 and *7.18* are audio of a gunshot with a high level of urban noise. The time shift moved the gunshot a bit later. After running the command repeatedly, Pluto found the time shift with a **minimum fraction** of 0.5 acceptable for the US8K real-world dataset.

The next audio augmentation technique we'll cover is time stretching.

Time stretching

The definition and key code lines for the `play_aug_time_stretch()` function are as follows:

```
# function definition
def play_aug_time_stretch(self, df,
  min_rate=0.2,
  max_rate=6.8,
  leave_length_unchanged=True,
  title='Time Stretch'):
# code snippet for time stretch
xtransform = audiomentations.TimeStretch(
  min_rate = min_rate,
  max_rate = max_rate,
  leave_length_unchanged = leave_length_unchanged,
  p=1.0)
```

The fill function's code can be found in the Python Notebook. Pluto tests the time stretch wrapper function with the audio control file and a **maximum rate** of 5 . 4, as follows:

```
# augment using time stretch
pluto.play_aug_time_stretch(pluto.audio_control_dmajor,
  max_rate=5.4)
```

The output for the time stretch augmented audio clip is as follows:

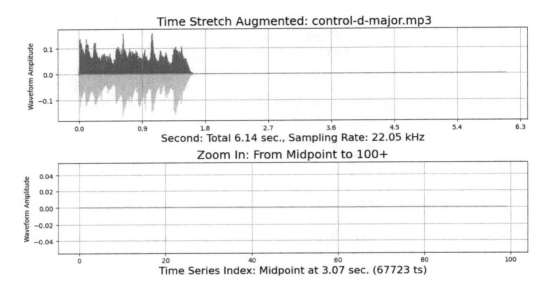

Figure 7.20 – Time stretch (control-d-major.mp3)

The wrapper function also displays the original audio clip for comparison:

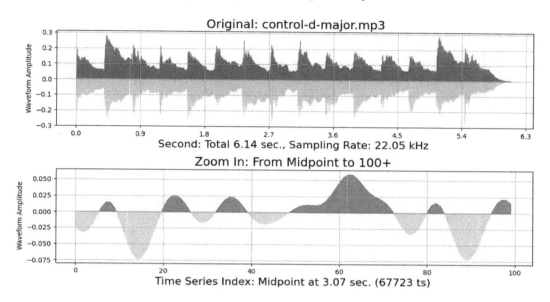

Figure 7.21 – Original time stretch (control-d-major.mp3)

Pluto runs the wrapper function repeatedly, and the scaled audio is recognizable every time. *Figure 7.20* audio plays the D major clip about three times faster, but the scales are recognizable.

The wrapper function works well on the control audio files, so Pluto applies to the music (MEC) dataset with a **maximum rate** of 3 . 0, as follows:

```
# augment using tim stretch
pluto.play_aug_time_stretch(pluto.df_music_data,
   max_rate=3.0)
```

The output for the time stretch augmented audio clip is as follows:

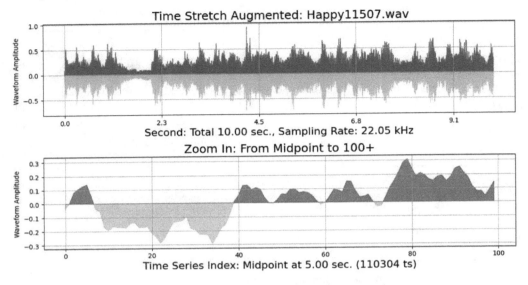

Figure 7.22 – Time stretch music (Sad44404.wav)

The wrapper function also displays the original audio clip for comparison:

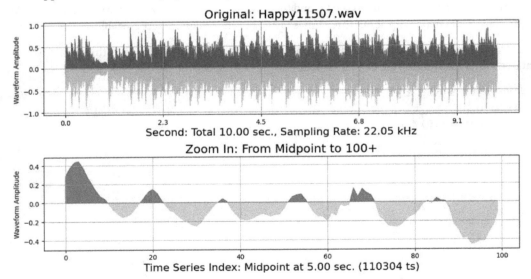

Figure 7.23 – Original time stretch music (Sad44404.wav)

The audio in *Figures 7.22* and *7.23* is of an afternoon lunch in a garden with a strong lead guitar and cinematic orchestra clip. With the time stretch filter at a **maximum rate** of 3 . 0, the audio in *Figure 7.22* plays a bit faster, but Pluto did not notice any degradation in the music's mood. Pluto repeatedly ran the wrapper function on the MEC dataset and concluded that the time stretch technique is **safe** at a **maximum rate** of 3 . 0.

> **Fun challenge**
>
> Find a universal **safe** range for the time stretch technique for all types of music (MEC). You can use the Python Notebook to find a safe range for the MEC datasets and download other music datasets from the Kaggle website. On the other hand, is this an impossible task? Does a universal safe range exist for pop, classical, folklore, country, and hip-hop music?

Pluto does the same for the human speech (CREMA-D) dataset. The command is as follows:

```
# augment using time stretch
pluto.play_aug_time_stretch(pluto.df_voice_data,
    max_rate=3.5)
```

The output for the time stretch augmented audio clip is as follows:

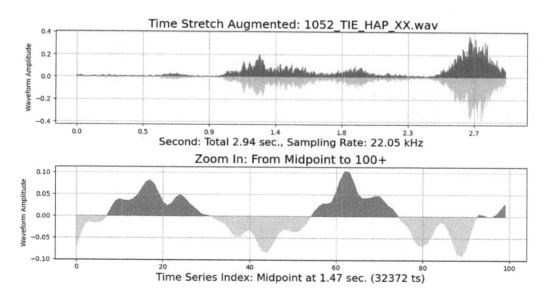

Figure 7.24 – Time stretch voice clip (1073_WSI_SAD_XX.wav)

The wrapper function also displays the original audio clip for comparison:

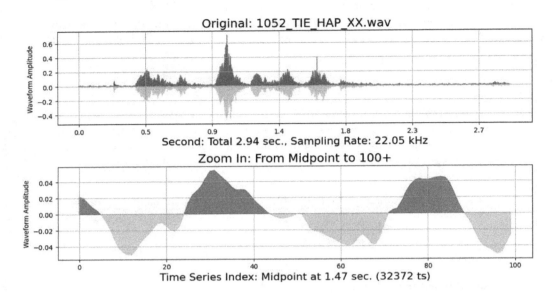

Figure 7.25 – Original time stretch voice clip (1073_WSI_SAD_XX.wav)

The audio in *Figures 7.24* and *7.25* is of a woman's voice saying, *Let's stop for a couple of minutes*, while the audio in *Figure 7.24* says it a bit faster but it's recognizable. Pluto repeatedly runs the wrapper function on the CREMA-D dataset with a **maximum rate** of 3 . 5 and hears no deterioration in the recordings. Thus, he concluded that the CREMA-D dataset is **safe** for use with the time stretch technique set to a **maximum rate** of 3 . 5.

Pluto does the same for the urban sound (US8K) dataset. The command is as follows:

```
# augment using time stretch
pluto.play_aug_time_stretch(pluto.df_sound_data,
    max_rate=2.4)
```

The output for the time stretch augmented audio clip is as follows:

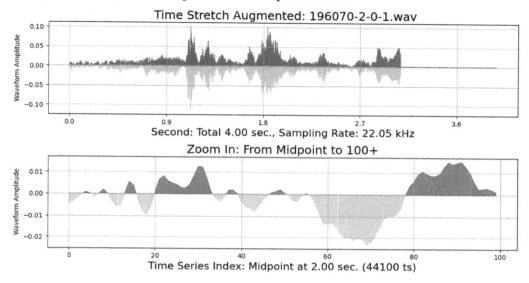

Figure 7.26 – Time stretch urban sound (76266-2-0-50.wav)

The wrapper function also displays the original audio clip for comparison:

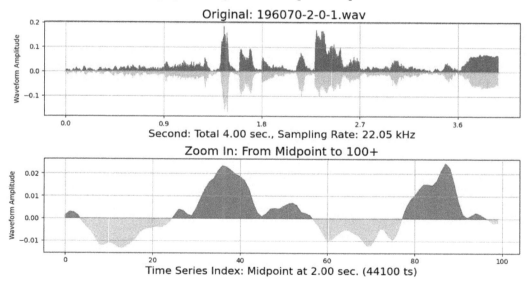

Figure 7.27 – Original time stretch urban sound (76266-2-0-50.wav)

The audio in *Figures 7.26* and *7.27* is of an urban clip of adults and children talking in a playground with high traffic or wind noises in the recording. The audio in *Figure 7.26* plays a bit faster. Pluto repeatedly runs the wrapper function on the US8K dataset with a **maximum rate** of 2.4, and he concluded that the US8K dataset is **safe** for the time stretching technique.

The next technique we'll look at is **pitch scaling**.

Pitch scaling

The definition and key code lines for the `play_aug_pitch_scaling()` function are as follows:

```
# function definition
def play_aug_pitch_scaling(self, df,
  min_semitones = -6.0,
  max_semitones = 6.0,
  title='Pitch Scaling'):
# code snippet for pitch shift
xtransform = audiomentations.PitchShift(
  min_semitones = min_semitones,
  max_semitones = max_semitones,
  p=1.0)
```

Pluto tests the pitch scaling wrapper function with the audio control file using the default parameters, as follows:

```
# augment using pitch scaling
pluto.play_aug_pitch_scaling(pluto.audio_control_dmajor)
```

The output augmented audio clip is as follows:

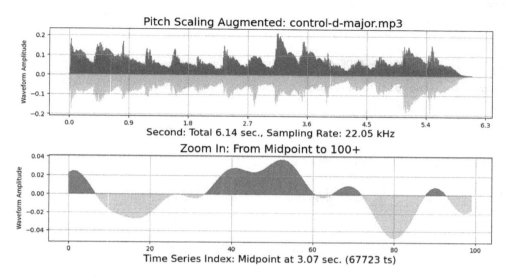

Figure 7.28 – Pitch scaling (control-d-major.mp3)

The wrapper function also displays the original audio clip for comparison:

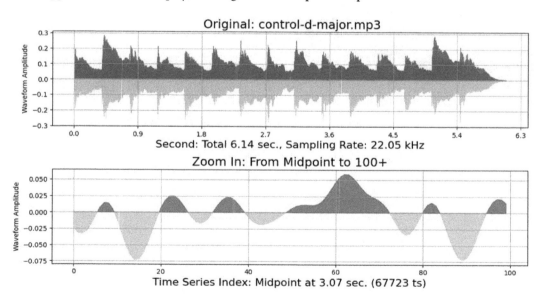

Figure 7.29 – Original pitch scaling (control-d-major.mp3)

Pluto can't tell the difference from looking at the complete Waveform graphs in *Figures 7.28* and *7.29*, but if he focuses on the zoom-in chart, he can see the differences. Listening to the audio file is the best method. To do that, you must go to the Python Notebook and click on the audio play button. The audio in *Figure 2.28* plays more like a harpsichord than the original piano scale in D major.

Next, Pluto applies the pitch scaling wrapper function to the music (MEC) dataset, as follows:

```
# augment using pitch scaling
pluto.play_aug_pitch_scaling(pluto.df_music_data,
  min_semitones=-11.0,
  max_semitones=-9.0)
```

The output for the augmented audio clip is as follows:

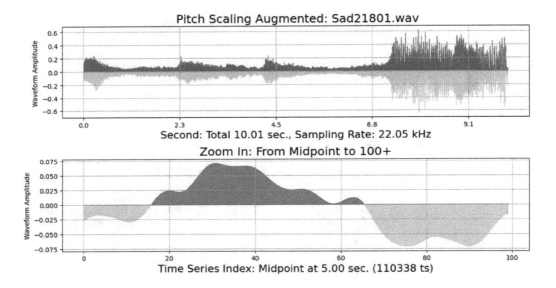

Figure 7.30 – Pitch scaling music (Sad11601.wav)

The wrapper function also displays the original audio clip for comparison:

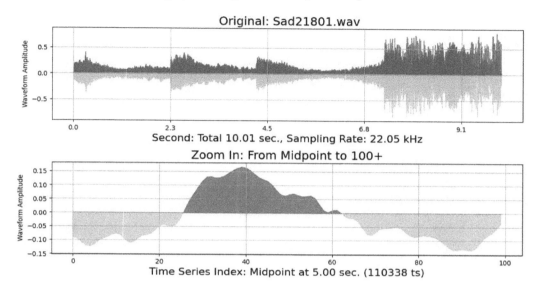

Figure 7.31 – Original pitch scaling music (Sad11601.wav)

The audio in *Figure 7.30* plays warmer, is melodic, and accentuates the moodiness of a dramatic cinematic clip. It's like a calm evening before a dramatic turn. Pluto purposefully exaggerated the pitch effects by setting the **minimum semitones** to -11.0 and the **maximum semitones** to -9.0. The audio in *Figure 7.31* plays the original clip. Using the default parameter value, Pluto found minimal pitch scaling effects on the MEC dataset. Thus, it is a **safe** technique.

Using the default parameter values, Pluto does the same for the voice (CREMA-D) dataset. The command is as follows:

```
# augment using pitch scaling
pluto.play_aug_pitch_scaling(pluto.df_voice_data)
```

The output for the augmented audio clip is as follows:

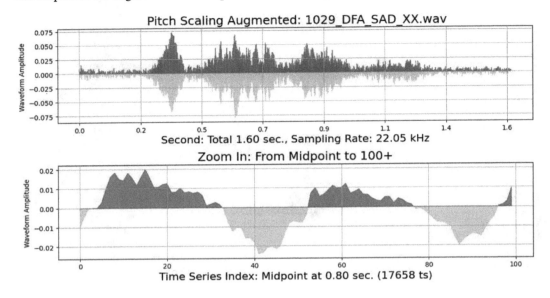

Figure 7.32 – Pitch scaling voice (1031_IEO_ANG_LO.wav)

The wrapper function also displays the original audio clip for comparison:

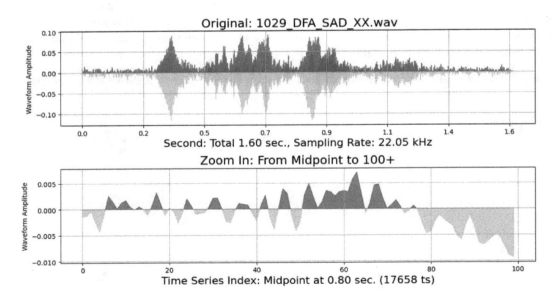

Figure 7.33 - Original pitch scaling voice (1031_IEO_ANG_LO.wav)

Pluto compares the zoom-in graphs in *Figures 7.32* and *7.33* to see the effects. When listening to the audio files, he heard the augmented version, from *Figure 7.32*, of a high-pitched kid saying, *It is eleven o'clock.* The original version is an adult man's voice. After repeatedly running the wrapper command with safe **minimum and maximum semitones** set to -2.4 and 2.4, Pluto found it minimized the effects for the CREMA-D dataset.

The urban sound (US8K) dataset has a diverse frequency range. Machine noises are repetitive low-frequency sounds, while sirens are high-frequency sounds. Pluto could not find a safe range unless he limited the **semitones'** scope to -1.2 and 1.0. For fun, Pluto has moved the **semitones** range to 4.0 and 14.0. The command is as follows:

```
# augment using pitch scaling
pluto.play_aug_pitch_scaling(pluto.df_sound_data,
  min_semitones=4.0,
  max_semitones=11.0)
```

The output for the augmented audio clip is as follows:

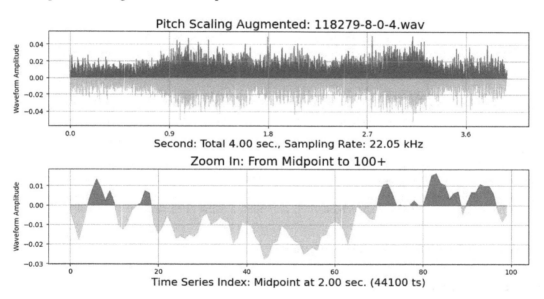

Figure 7.34 – Pitch scaling urban sound (93567-8-3-0.wav)

The wrapper function also displays the original audio clip for comparison:

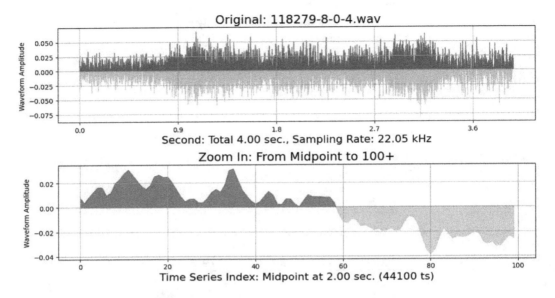

Figure 7.35 – Original pitch scaling urban sound (93567-8-3-0.wav)

The audio in *Figures 7.34* and *7.35* play an urban clip of sirens in a busy urban street. The audio in *Figure 7.34* has the sirens sound clearer and with a bit less interference from the traffic noise.

> **Fun challenge**
>
> This challenge is a thought experiment. Can you define rules for which audio augmentation methods are suitable for an audio category, such as human speech, music, bird songs, and so on? For example, can human speech be safely augmented using pitch shifting in a small range?

The next technique we'll look at is **noise injection**.

Noise injection

The definition and key code lines for the play_aug_noise_injection() function are as follows:

```
# function definition
def play_aug_noise_injection(self, df,
  min_amplitude = 0.002,
  max_amplitude = 0.2,
  title='Gaussian noise injection'):
# code snippet for noise injection
```

```
xtransform = audiomentations.AddGaussianNoise(
    min_amplitude = min_amplitude,
    max_amplitude = max_amplitude,
    p=1.0)
```

The full function's code can be found in the Python Notebook. Pluto will not explain the result here because they are similar to the previous three audio augmentation techniques. You should try them out on the Python Notebook to see and hear the results.

For the **background noise injection** method, the code snippet is as follows:

```
# code snippet for adding background noise
xtransform = audiomentations.AddBackgroundNoise(
    sounds_path="~/background_sound_files",
    min_snr_in_db=3.0,
    max_snr_in_db=30.0,
    noise_transform=PolarityInversion(),
    p=1.0)
```

For the **short noise injection** method, the code snippet is as follows:

```
# code snippet for adding short noise
xtransform = audiomentations.AddShortNoises(
    sounds_path="~/background_sound_files",
    min_snr_in_db=3.0,
    max_snr_in_db=30.0,
    noise_rms="relative_to_whole_input",
    min_time_between_sounds=2.0,
    max_time_between_sounds=8.0,
    noise_transform=audiomentations.PolarityInversion(),
    p=1.0)
```

The full function code can be found in the Python Notebook. The next technique we'll look at is **polarity inversion**.

Polarity inversion

The definition and key code lines for the `play_aug_polar_inverse()` function are as follows:

```
# function definition
def play_aug_polar_inverse(self, df,
    title='Polarity inversion'):
# code snippet for polarity inversion
xtransform = audiomentations.PolarityInversion(
    p=1.0)
```

Once again, Pluto will not explain the result here because they have similar outputs to what you saw previously. Try them out on the Python Notebook to see and hear the results. Pluto has written the Python code for you.

> **Fun fact**
>
> There is one fun fact about the polarity inversion technique: you will not hear any difference between the augmented and original recordings. You couldn't even see the difference in the WaveForm graph, but you could see it in the zoom-in chart. The blue/positive and yellow/negative are flipped.

For example, Pluto applies the wrapper function to the voice (CREMA-D) dataset as follows:

```
# augment using polar inverse
pluto.play_aug_polar_inverse(pluto.df_voice_data)
```

The output for the augmented audio clip is as follows:

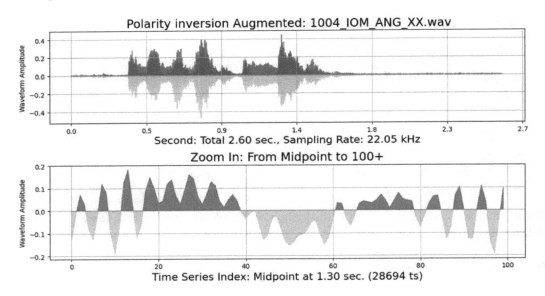

Figure 7.36 – Polar inversion voice (1081_WSI_HAP_XX.wav)

The wrapper function also displays the original audio clip for comparison:

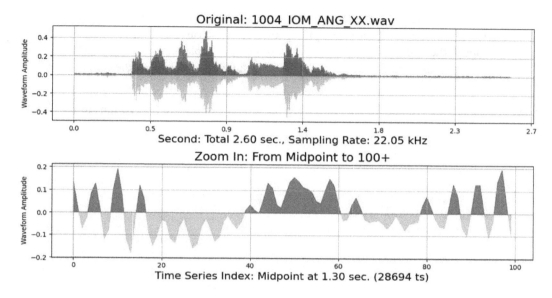

Figure 7.37 – Original polar inversion voice (1081_WSI_HAP_XX.wav)

Another fun fact is that polar inversion is as simple as multiplying the **amplitude** array with a negative one, like so:

```
# implement using numpy
xaug = numpy.array(data_amp) * -1
```

> **Fun challenge**
>
> Here is a thought experiment: why does polarity inversion not affect the sound? After all, it is a drastic change in the data, as evidenced in the Waveform graph. Hint: think about the technical complexities of molecules' vibration from compression and expansion relating to absolute measurement.

The next few techniques we'll look at use **filters**.

Low-pass filter

Before Pluto digs in and explains the filter's audio techniques, he will only partially present all the filters in this book. This is because the process is repetitive, and you can gain much more insight from running the code in the Python Notebook. Pluto will thoroughly explain the code and the Waveform graph for the **low-pass** and **band-pass** filters; for the other filters, he will explain the code but not the output Waveform graphs.

The definition and key code lines for the `play_aug_low_pass_filter()` function are as follows:

```
# function definition
def play_aug_low_pass_filter(self, df,
  min_cutoff_freq=150, max_cutoff_freq=7500,
  min_rolloff=12, max_rolloff=24,
  title='Low pass filter'):
# code snippet for low pass filter
xtransform = audiomentations.LowPassFilter(
  min_cutoff_freq = min_cutoff_freq,
  max_cutoff_freq = max_cutoff_freq,
  min_rolloff = min_rolloff,
  max_rolloff = max_rolloff,
  p=1.0)
```

The full function's code can be found in the Python Notebook. Pluto tests the low-pass filter wrapper function with the audio control file using default parameters, as follows:

```
# augment using low pass filter
pluto.play_aug_low_pass_filter(pluto.audio_control_dmajor)
```

The output for the augmented audio clip is as follows:

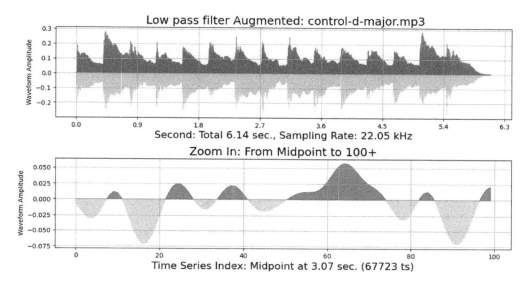

Figure 7.38 – Low-pass filter control (control-d-major.mp3)

The wrapper function also displays the original audio clip for comparison:

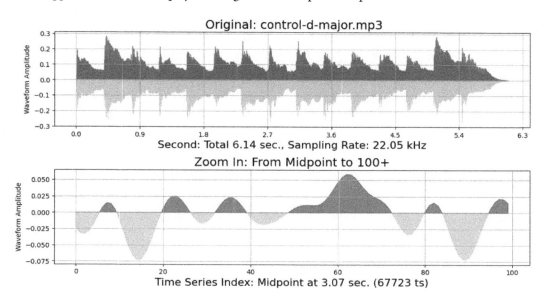

Figure 7.39 – Original low-pass filter control (control-d-major.mp3)

Pluto does not detect any difference in listening to the augmented and original recordings shown in *Figures 7.38* and *7.39*. At first glance at the WaveForm graphs, Pluto did not see any differences until he inspected the zoom-in charts. There is a slight decrease in the positive (blue color) **amplitude** values and, inversely, a tiny increase in the negative (yellow color) values. In other words, the absolute differences between the before and after are slightly lower **amplitude** values.

Next, Pluto applies the low-pass filter wrapper function to the music (MEC) dataset, as follows:

```
# augment using low pass filter
pluto.play_aug_low_pass_filter(pluto.df_music_data)
```

The output for the augmented audio clip is as follows:

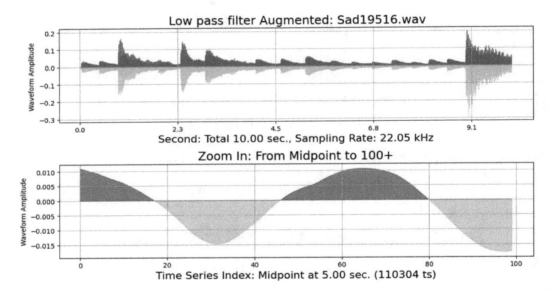

Figure 7.40 – Low-pass filter music (Sad21828.wav)

The wrapper function also displays the original audio clip for comparison:

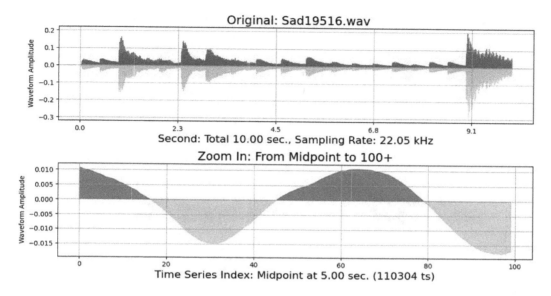

Figure 7.41 – Original low-pass filter music (Sad21828.wav)

The audio in *Figures 7.40* and *7.41* is of a cinematic orchestra music clip with a driving drum beat. It could be the background music from an Indiana Jones movie before the giant boulder bars down the cave. In particular, the augmented file sounds, shown in *Figure 7.40*, are smoother and the edges have been nipped off. Pluto repeatedly ran the wrapper function on the MEC dataset using the default parameter settings and found that the augmented audio file does not alter the happy or sad mood of the music. Hence, it is a **safe** technique.

For the voice (CREMA-D) dataset, Pluto does the same:

```
# augment using low pass filter
pluto.play_aug_low_pass_filter(pluto.df_voice_data)
```

The output for the augmented audio clip is as follows:

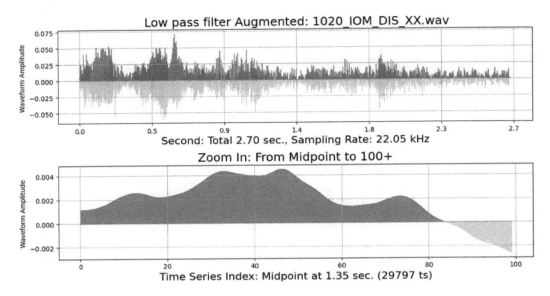

Figure 7.42 – Low-pass filter voice (1067_IEO_HAP_LO.wav)

The wrapper function also displays the original audio clip for comparison:

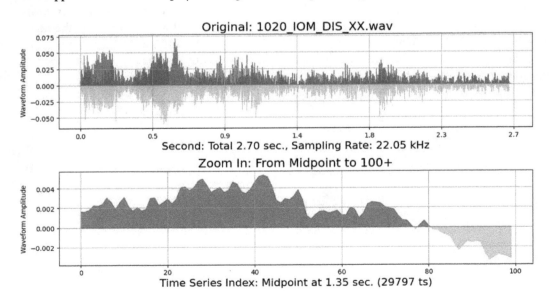

Figure 7.43 – Original low-pass filter voice (1067_IEO_HAP_LO.wav)

The audio in *Figures 7.42* and *7.43* both said *It is eleven o'clock*. Furthermore, the audio in *Figure 7.42* has fewer snaps and crackles. Pluto has an unscientific thought that the zoom-in graph displays a smoother curve with fewer dips and dimples, which could translate to a cleaner voice in the augmented recording. After repeatedly applying the wrapper function to the CREMA-D dataset, Pluto thinks the low-pass filter is **safe**.

The last of the three real-world datasets is the urban sound (US8K) dataset. Pluto applies the wrapper function as follows:

```
# augment using low pass filter
pluto.play_aug_low_pass_filter(pluto.df_sound_data)
```

The output for the augmented audio clip is as follows:

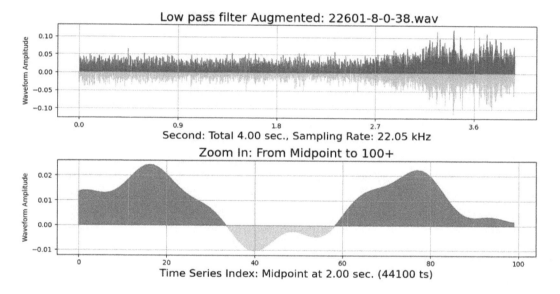

Figure 7.44 – Low-pass filter urban sound (185373-9-0-6.wav)

The wrapper function also displays the original audio clip for comparison:

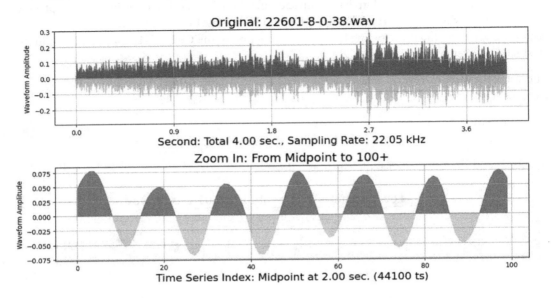

Figure 7.45 – Original low-pass filter urban sound (185373-9-0-6.wav)

The audio in *Figures 7.44* and *7.45* is of street music playing outdoors with traffic and urban noise. Repeatedly executing the wrapper functions gives mixed results for the US8K dataset. Pluto does not know which parameter values are **safe**. He needs to consult a domain expert – that is, a sound engineer.

The next technique we'll look at is the **band-pass filter**.

Band-pass filter

The definition and key code lines for the `play_aug_band_pass_filter()` function are as follows:

```
# function definition
def play_aug_band_pass_filter(self, df,
  min_center_freq=200, max_center_freq=4000,
  min_bandwidth_fraction=0.5, max_bandwidth_fraction=1.99,
  min_rolloff=12, max_rolloff=24,
  title='Band pass filter'):
# code snippet for band pass filter
xtransform = audiomentations.BandPassFilter(
  min_center_freq = min_center_freq,
  max_center_freq = max_center_freq,
  min_bandwidth_fraction = min_bandwidth_fraction,
  max_bandwidth_fraction = max_bandwidth_fraction,
```

```
min_rolloff = min_rolloff,
max_rolloff = max_rolloff,
p=1.0)
```

The full function's code can be found in the Python Notebook. Pluto tests the band-pass filter wrapper function with the audio control file using default parameters, as follows:

```
# augment using band pass filter
pluto.play_aug_band_pass_filter(pluto.audio_control_dmajor)
```

The output for the augmented audio clip is as follows:

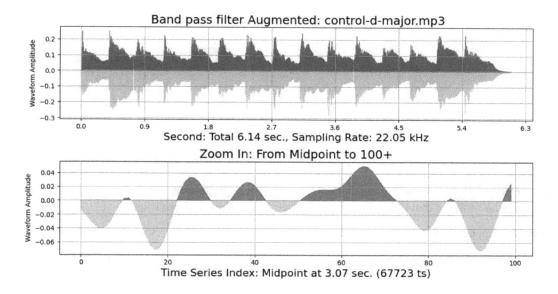

Figure 7.46 – Band-pass filter control (control-d-major.mp3)

The wrapper function also displays the original audio clip for comparison:

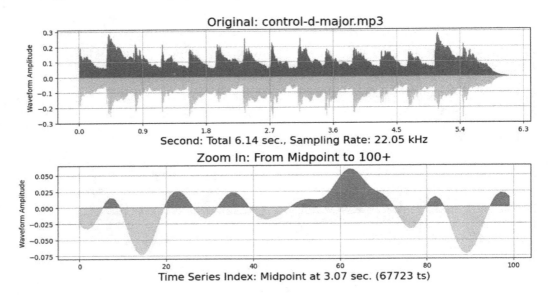

Figure 7.47 – Original band-pass filter control (control-d-major.mp3)

From *Figure 7.46*, Pluto could guess that the sound has been slightly altered. When listening to the audio file, he confirms that the scale is the same, but it has a whom-whom sound to it.

Next, Pluto applies the band-pass filter function to the music (MEC) dataset. The command is as follows:

```
# augment using band pass filter
pluto.play_aug_band_pass_filter(pluto.df_music_data)
```

The output for the augmented audio clip is as follows:

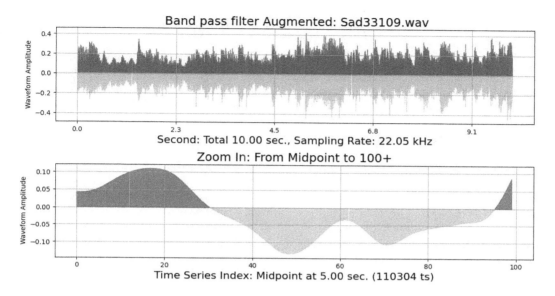

Figure 7.48 – Band-pass filter music (Happy15804.wav)

The wrapper function also displays the original audio clip for comparison:

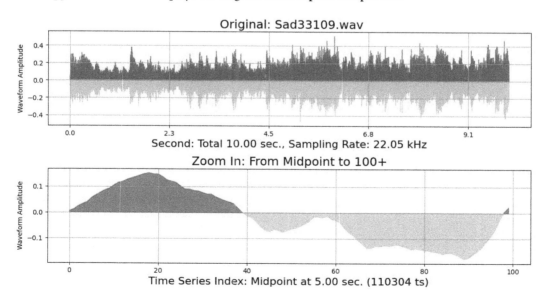

Figure 7.49 – Original band-pass filter music (Happy15804.wav)

The sound for this clip is a happy-go-lucky cinematic tune with a sprinkle of a drum beat. The augmented sound, shown in *Figure 7.48*, is brighter, bunchier, and yet smoother. Pluto repeatedly executes the wrapper function against the **MEC** dataset, and it enhances the **happier** mood music and infuses a more substantial tone into the **sadder** clips. Thus, it is **safe** for the MEC dataset.

Pluto does the same for the voice (CREMA-D) dataset. The command is as follows:

```
# augment using band pass filter
pluto.play_aug_band_pass_filter(pluto.df_voice_data)
```

The output for the augmented audio clip is as follows:

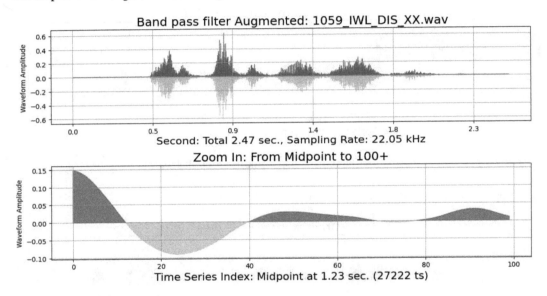

Figure 7.50 – Band-pass filter voice (1071_IWL_NEU_XX.wav)

The wrapper function also displays the original audio clip for comparison:

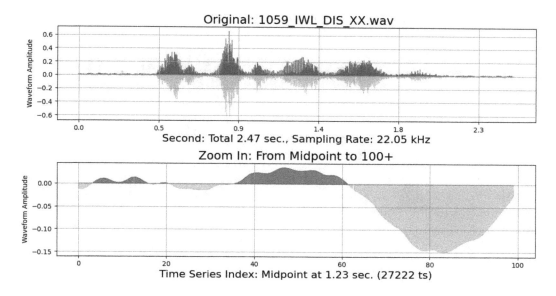

Figure 7.51 – Original band-pass filter voice (1071_IWL_NEU_XX.wav)

The audio for *Figures 7.50* and *7.51* is of a woman saying, *I would like a new alarm clock*. The augmented audio file sounds cleaner with less noise interference than the original clip. The same results were found for most of the files in the CREMA-D dataset. Thus, the CREMA-D dataset is **safe** for use with the band-pass filter technique.

Pluto suspects the same improvement or at least a safe level for the urban sound (US8K) dataset. The command is as follows:

```
# augment using band pass filter
pluto.play_aug_band_pass_filter(pluto.df_sound_data)
```

The output for the augmented audio clip is as follows:

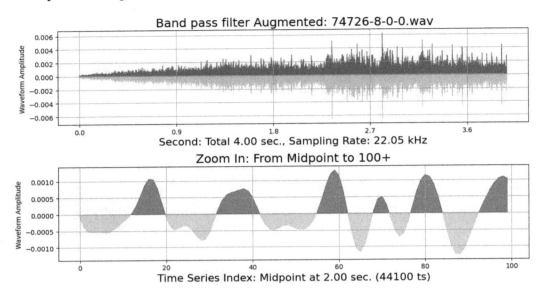

Figure 7.52 – Band-pass filter urban sound (95404-3-0-0.wav)

The wrapper function also displays the original audio clip for comparison:

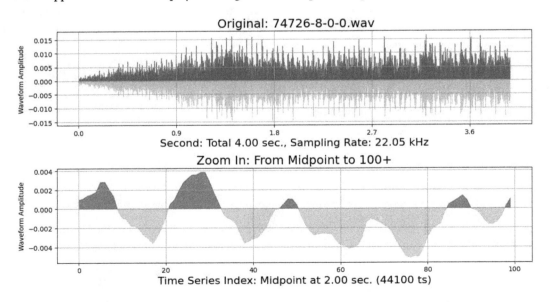

Figure 7.53 – Original band-pass filter urban sound (95404-3-0-0.wav)

The audio file for this is the sound of a windy backyard with birds singing and fading dogs barking from far away. The augmented audio file, *Figure 7.52*, sounds more distinct but with echoes in a tunnel effect. Pluto thinks the band-pass filter is **safe** for the US8K dataset.

> **Fun challenge**
>
> Pluto challenges you to implement the reversed audio technique. Can you think of a use case where reversed audio is a **safe** option? Hint: copy and rename the `play_aug_time_shift()` wrapper function. Change `xtransform = audiomentations.Shift()` to `xtransform = audiomentations.Reverse()`.

The audio augmentation process becomes slightly repetitive, but the results are fascinating. Thus, Pluto has shared the code for the following audio filter techniques, but the resulting WaveForm graphs and audio play buttons are in the Python Notebook.

The next filter we'll cover is the **high-pass filter**.

High-pass and other filters

The definition and key code lines for the `play_aug_high_pass_filter()` function are as follows:

```
# function definition
def play_aug_high_pass_filter(self, df,
  min_cutoff_freq=20, max_cutoff_freq=2400,
  min_rolloff=12, max_rolloff=24,
  title='High pass filter'):
# code snippet for high pass filter
xtransform = audiomentations.HighPassFilter(
  min_cutoff_freq = min_cutoff_freq,
  max_cutoff_freq = max_cutoff_freq,
  min_rolloff = min_rolloff,
  max_rolloff = max_rolloff,
  p=1.0)
```

The results can be found in the Python Notebook.

> **Fun challenge**
>
> Pluto challenges you to implement other audio filters in the Audiomentations library, such as `audiomentations.HighPassFilter`, `audiomentations.LowShelfFilter`, `audiomentations.HighShelfFilter`, `audiomentations.BandStopFilter`, and `audiomentations.PeakingFilter`.

With that, we have covered the fundamentals of audio augmentations and practiced coding them. Next, we will summarize this chapter.

Summary

As we saw from the beginning, audio augmentation is a challenging topic without hearing the audio recording in question, but we can visualize the techniques' effect using **Waveform** graphs and zoom-in charts. Still, there is no substitution for listening to the before and after augmentation recordings. You have access to the Python Notebook with the complete code and audio button to play the augmented and original recordings.

First, we discussed the theories and concepts of an audio file. The three fundamental components of an audio file are **amplitude, frequency**, and **sampling rate**. The measurements of unit for frequency are **Hertz (Hz)** and **Kilohertz (kHz)**. **Pitch** is similar to frequency, but the unit of measurement is the **decibel (dB)**. Similarly, **bit rate** and **bit depth** are other forms expressing the sampling rate.

Next, we explained the standard audio augmentation techniques. The three essentials are **time stretching, time shifting**, and **pitch scaling**. The others are **noise injection** and **polarity inversion**. Even more methods are available in the augmentation libraries, such as clip, gain, normalize, and **hyperbolic tangent (tanh)** distortion.

Before downloading the real-world audio datasets, we discussed the top eight open source audio augmentation libraries. There are many robust audio augmentation libraries available. Pluto picked the **Librosa** library – after all, it's the most established. Its second choice was the **Audiomentations** library because it is powerful and easy to integrate with other libraries. Facebook's **Augly** libraries are strong contenders, and Pluto used them in other projects. Ultimately, because Pluto uses the wrapper function concept, he can choose any library or combination of libraries.

As with image and text augmentation, Pluto downloaded three real-world audio datasets from the *Kaggle* website. Each dataset represents an audio category in everyday experiences: music, human speech, and urban sound.

Writing code in the Python Notebook helps us reinforce our understanding of each audio augmentation technique. Pluto explains the code and the output in detail.

The output is fantastic, but the coding process seems repetitive. It is easy because Pluto follows the established pattern of creating a reusable class, adding new methods, downloading real-world data from the *Kaggle* website, importing it into pandas, leveraging best-of-class augmentation libraries, and writing new wrapper functions.

Throughout this chapter, there were *fun facts* and *fun challenges*. Pluto hopes you will take advantage of these and expand your experience beyond the scope of this chapter.

In the next chapter, Pluto will demystify audio using **spectograms**.

8
Audio Data Augmentation with Spectrogram

In the previous chapter, we visualized the sound using the Waveform graph. An audio spectrogram is another visualizing method for seeing the audio components. The inputs to the Spectrogram are a one-dimensional array of **amplitude** values and the **sampling rate**. They are the same inputs as the Waveform graph.

An audio **spectrogram** is sometimes called a **sonograph, sonogram, voiceprint,** or **voicegram**. The Spectrogram is a more detailed representation of sound than the Waveform graph. It shows a correlation between frequency and amplitude (loudness) over time, which helps visualize the frequency content in a signal. Spectrograms make it easier to identify musical elements, detect melodic patterns, recognize frequency-based effects, and compare the results of different volume settings. Additionally, the Spectrogram can be more helpful in identifying non-musical aspects of a signal, such as noise and interference from other frequencies.

The typical usage is for music, human speech, and sonar. A short standard definition is a spectrum of frequency maps with time duration. In other words, the y axis is the frequency in **Hz or kHz**, and the x axis is the time duration in **seconds or milliseconds**. Sometimes, the graph comes with a color index for the amplitude level.

Pluto will explain the code in the Python Notebook later in the chapter, but here is a sneak peek of an audio Spectrogram. The command for drawing the *control piano scale in the D major* audio file is as follows:

```
# draw Spectrogram
pluto.draw_spectrogram(pluto.audio_control_dmajor)
```

The output is as follows:

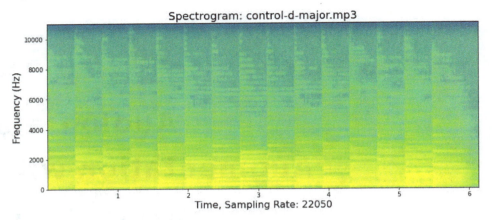

Figure 8.1 – An audio spectrogram of piano scale in D major

Before Pluto demystifies the audio Spectrogram, you should review *Chapter 7* if the audio concepts and keywords sound alien to you. This chapter relies heavily on the knowledge and practices from *Chapter 7*.

In *Figure 8.1*, Pluto uses the Matplotlib library to draw the audio spectrograph. The primary input is the amplitude array and the sampling rate. The library does all the heavy calculations, and other libraries, such as the Librosa or SciPy library, can perform the same task. In particular, Matplotlib can generate many types of audio spectrographs from the same input. Pluto will dig deeper into types of spectrographs a bit later, but first, let's break down the steps of how the library constructs an audio spectrograph. The five high-level tasks are as follows:

1. Splitting the audio stream into overlapping segments, also known as **windows**.

2. Calculating the **Short-Time Fourier Transform (STFT)** value on each window.

3. Converting the windows' value into **decibels (dB)**.

4. Linking the windows together as in the original audio sequence.

5. Displaying the result in a graph with the *y* axis as Hz, the *x* axis as seconds, and dB as a color-coded value.

The math for the previous five steps is complex, and the chapter's goal is to use a Spectrogram to visualize the sound and augment the audio file. Thus, we rely on audio libraries to perform the math calculation.

As mentioned, the underlying data representing the Spectrogram is the same as the Waveform format. Therefore, the audio augmentation techniques are the same. Consequently, the resulting augmented audio file will sound the same. The visual representation is the only difference between the Spectrogram and the Waveform graph.

The majority of this chapter will cover the audio Spectrogram standard format, a variation of a Spectrogram, **Mel-spectrogram**, and **Chroma** STFT. The augmentation techniques represent a shorter section because you have learned the method in the previous chapter. We will cover the following topics in this chapter:

- Initializing and downloading

- Audio Spectrogram

- Various Spectrogram formats

- Mel-spectrogram and Chroma STFT plots

- Spectrogram augmentation

- Spectrogram image

> **Fun fact**
>
> The **Kay Electric Company** introduced the first commercially available machine for audio spectrographic analysis in 1951. The black-and-white image was named a sonograph or sonogram for visualizing bird songs. In 1966, **St. Martin's Press** used sonography for the book *Golden Field Guide to Birds of North America*. Spectrograms were favored over sonogram terminology around 1995 during the digital age. Spectrograms or sonograms were not limited to the study of birds in the early days. The US military used Spectrogram for encryption in the early 1940s and continues forward, as evidenced by the publication *Cryptologic Quarterly*, volume 38, published by the **Center for Cryptologic History** in 2019.

This chapter reuses the audio augmentation functions and the real-world audio datasets from *Chapter 7*. Thus, we will start by initializing Pluto and downloading the real-world datasets.

Initializing and downloading

Start with loading the `data_augmentation_with_python_chapter_8.ipynb` file on Google Colab or your chosen Python Notebook or JupyterLab environment. From this point onward, the code snippets are from the Python Notebook, which contains the complete functions.

The following initializing and downloading steps should be familiar to you because we have done them six times. The following code snippet is the same as from *Chapter 7*:

```
# Clone GitHub repo.
url = 'https://github.com/PacktPublishing/Data-Augmentation-with-
Python'
!git clone {url}
# Intialize Pluto from Chapter 7
pluto_file = 'Data-Augmentation-with-Python/pluto/pluto_chapter_7.py'
%run {pluto_file}
```

```
# Verify Pluto
pluto.say_sys_info()
# Fetch Musical emotions classification
url = 'https://www.kaggle.com/datasets/kingofarmy/musical-emotions-
classification'
pluto.fetch_kaggle_dataset(url)
f = 'kaggle/musical-emotions-classification/Train.csv'
pluto.df_music_data = pluto.fetch_df(f)
# Fetch human speaking
url = 'https://www.kaggle.com/datasets/ejlok1/cremad'
pluto.fetch_kaggle_dataset(url)
f = 'kaggle/cremad/AudioWAV'
pluto.df_voice_data = pluto.make_dir_dataframe(f)
# Fetch urban sound
url='https://www.kaggle.com/datasets/rupakroy/urban-sound-8k'
pluto.fetch_kaggle_dataset(url)
f = 'kaggle/urban-sound-8k/UrbanSound8K/UrbanSound8K/audio'
pluto.df_sound_data = pluto.make_dir_dataframe(f)
```

> **Fun challenge**
>
> Pluto challenges you to search for and download an additional real-world audio dataset from the *Kaggle* website or your project. A hint is to use Pluto's `fetch_kaggle_data()` and `fetch_df()` methods, and any of the audio augmentation wrapper functions.

A few under-the-hood methods make the process so easy to use. Pluto highly recommends that you review *Chapter 7* before continuing with the spectrogram.

Audio Spectrogram

Before dissecting the Spectrogram, let's review the fundamental differences between a Spectrogram and a Waveform plot. The Spectrogram graphs show the frequency components of a sound signal over time, focusing on frequency and intensity. In contrast, the Waveforms concentrate on the timing and amplitude of sounds. The difference is in the visual representation of the sound wave. The underlying data representation and the transformation methods are the same.

An audio Spectrogram is another visual representation of a sound wave, and you saw the Waveform graph in *Chapter 7*. The `_draw_spectrogram()` helper method uses the Librosa library to import the audio file and convert it into an amplitude data one-dimensional array and a sampling rate in Hz. The next step is to use the Matplotlib library to draw the Spectrogram plot. Likewise, Pluto takes the output from the Librosa library function and uses the Matplotlib function to draw the fancy blue and yellow Waveform graph in *Chapter 7*. The relevant code snippet is as follows:

```
# read audio file
data_amp, sam_rate = librosa.load(lname, mono=True)
```

```
# draw the spectrogram plot
spectrum, freq, ts, ax = pic.specgram(data_amp, Fs=sam_rate)
```

Here, the returned values are as follows:

- `spectrum` is a `numpy.array` type with `shape(n,m)`. For example, the result of plotting the Spectrogram of the *control piano scale in a D major* audio file `shape()` is equal to `(129, 1057)`. It represents the m-column of periodograms for each segment or window.

- `freq` is a `numpy.array` type with `shape(n,)`. Using the same example, `freq shape` is `(129,)`. It represents the frequencies corresponding to the elements (rows) in the `spectrum` array.

- `ts` is a `numpy.array` type with `shape(n,)`. Using the same example as previously, `ts shape` is `(1057,)`. It represents the times corresponding to midpoints of `spectrum`'s n-column.

- `ax` is a `matplotlib.image.AxesImage` type. It is the image from the Matplotlib library.

Pluto draws the Spectrogram for the control piano scale in D major audio file using the following command:

```
# plot the Spectrogram
pluto.draw_spectrogram(pluto.audio_control_dmajor)
```

The output is as follows:

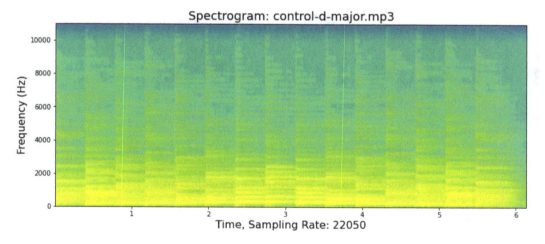

Figure 8.2 – An audio Spectrogram of piano scale in D-major

Pluto displays the audio-play button in the Python Notebook, where you can listen to the audio. The button image looks like the following:

Figure 8.3 – The audio play button

For comparison, the following is the Waveform graph from *Chapter 7* using the helper function:

```
# plot the Waveform
pluto._draw_audio(data_amp, sam_rate, 'Original: ' + fname)
```

The output is as follows:

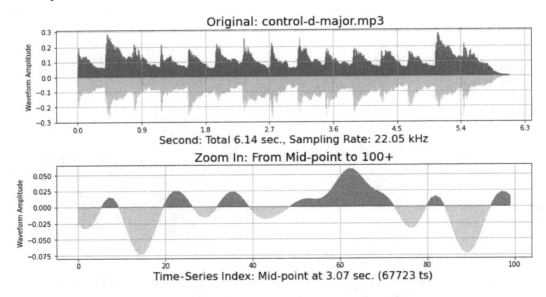

Figure 8.4 – Audio waveform of piano scale in D major

The music sounds the same. Only the visual displays are different.

Sound engineers are trained to read Spectrogram plots to identify and remove unwanted noises, such as the following:

- **Hum**: This is usually the result of electrical noise in the recording. Its range is typically between 50 Hz and 60 Hz.

- **Buzz**: This is the opposite of hum. It is the electrical noise of higher frequencies. Familiar sources are fluorescent light fixtures, on-camera microphones, and high-pitched motors.

- **Hiss**: This is a broadband noise, which is different from hum and buzz. It is typically concentrated at specific frequencies in both upper and lower spectrums. The usual suspects are **heating, ventilation, and air conditioning (HVAC)** systems or motor fans.

- **Intermittent noises**: These are commonly introduced by urban sounds such as thunders, birds, wind gusts, sirens, car horns, footsteps, knocking, coughs, or ringing cell phones.

- **Digital clipping**: This is when the audio is too loud to be recorded. It is the loss of the audio signal's peaks.

- **Gaps**: Gaps or dropouts are silences due to missing cut-outs in the audio recording.

- **Clicks and pops**: These are noises in the recording caused by vinyl and other grooved media recording devices.

Pluto uses the **Matplotlib** library function, which has many parameters governing the display of the Spectrogram plots. Let's use the three real-world audio datasets to illustrate other visual representations of Spectrogram plots.

Various Spectrogram formats

There are many parameters Pluto can pass to the underlying `specgram()` method from the Matplotlib library. He will highlight only a few parameters.

> **Fun fact**
>
> You can print any function documentation by adding a question mark (?) at the end of the function in the Python Notebook.

For example, printing the documentation for the specgram() function is the following command: matplotlib.pyplot.specgram? The partial output is as follows:

Help ✕

```
Signature: pic.specgram(x, NFFT=None, Fs=None, Fc=None, detrend=None,
window=None, noverlap=None, cmap=None, xextent=None, pad_to=None,
sides=None, scale_by_freq=None, mode=None, scale=None, vmin=None,
vmax=None, *, data=None, **kwargs)
Docstring:
Plot a spectrogram.

Compute and plot a spectrogram of data in *x*.  Data are split into
*NFFT* length segments and the spectrum of each section is
computed.  The windowing function *window* is applied to each
segment, and the amount of overlap of each segment is
specified with *noverlap*. The spectrogram is plotted as a colormap
(using imshow).

Parameters
----------
x : 1-D array or sequence
    Array or sequence containing the data.

Fs : scalar
    The sampling frequency (samples per time unit).  It is used
    to calculate the Fourier frequencies, freqs, in cycles per time
    unit. The default value is 2.
```

Figure 8.5 – Partial print definition of specgram()

You can view the complete output of *Figure 8.5* in the Python Notebook. Another example is printing Pluto's draw_spectrogram() function documentation as follows: pluto.draw_spectrogram?.

The output is as follows:

Help   Help ✕

```
Signature: pluto.draw_spectrogram(df, fname='Spectrogram', window=<function
window_hanning at 0x7fe93bb7edc0>, cmap='viridis', sides='default',
mode='default')
Docstring: <no docstring>
File:      /content/<ipython-input-27-4acfc0b43527>
Type:      method
```

Figure 8.6 – The print definition of draw_spectrogram()

From *Figure 8.5*, the simple one is changing the color map (`cmap`) variable. There are more than 60 color maps in the Matplotlib library. Thus, Pluto will choose a different `cmap` color for each audio dataset. Sound engineers may use different color maps to highlight specific frequency properties for spotting patterns or noises. Changing the visual representation does not affect the sound quality or the data. Thus, selecting the color map based solely on your preferences is acceptable. If vivid pink and blue are your favorite, choose the `cool` cmap value.

The Spectrogram code for the music dataset is as follows:

```
# plot the spectrogram in different color map
pluto.draw_spectrogram(pluto.df_music_data, cmap='plasma')
```

The output is as follows:

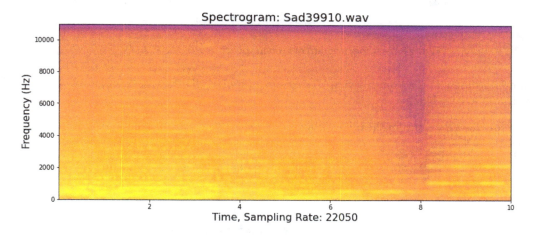

Figure 8.7 – Spectrogram of a music file (Sad39910)

Every time Pluto runs the `draw_spectrogram()` wrapper function, a random audio file is selected from the dataset. *Figure 8.7* is the audio of cinematic music with strong cello leads, and the `plasma` color map is a bright yellow transit to orange and deep blue-purple.

Likewise, for the human speech dataset, the command is as follows:

```
# plot the Spectrogram in different color map
pluto.draw_spectrogram(pluto.df_voice_data, cmap='cool')
```

The output is as follows:

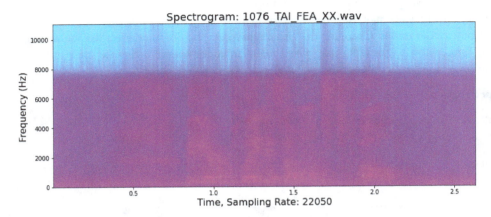

Figure 8.8 – A spectrogram of human speech (1076_TAI_FEA_XX)

Figure 8.8 is the audio of a woman saying "*The airplane is almost full*". The cool color map is a fuchsia pink transit to baby blue.

Next, Pluto does the same for the urban sound dataset using the following command:

```
# plot the Spectrogram with different color map
pluto.draw_spectrogram(pluto.df_sound_data, cmap='brg')
```

The output is as follows:

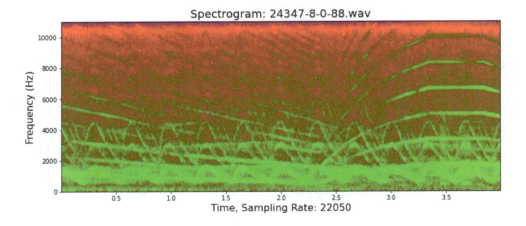

Figure 8.9 – A spectrogram of urban sound (24347-8-0-88)

Figure 8.9 sounds like a passing siren from an ambulance. The `brg` color map is blue, red, and green, making a striking and dramatic graph.

Fun challenge

The challenge is a thought experiment. Is a particular color map with a multicolor such as a `rainbow` cmap or two colors such as `ocean` cmap more advantageous for different types of audio such as urban sound or music? In other words, is displaying the Spectrogram for a human singing an audio clip better in pink and magenta shades or multicolor earth tones?

In audio engineering, a `window_hanning` parameter uses weighted cosine to diminish the audio spectrum. Window-hanning is a technique used to reduce artifacts in the frequency domain of an audio signal. It uses a `window` function to gently taper off the signal's amplitude near its edges, minimizing the effect of spectral leakage and reducing unwanted noise in the signal. Window-hanning also improves the time-domain resolution of the signal, making it easier to identify onsets and offsets with greater precision.

Pluto's `draw_spectrogram()` method uses it as the default value. What if Pluto wants to see the raw signal without `window_hanning`? He can use `window_none` on the control and voice dataset, as per the following command:

```
# control audio file
pluto.draw_spectrogram(pluto.audio_control_dmajor,
    window=matplotlib.mlab.window_none)
# Human speech
pluto.draw_spectrogram(pluto.df_voice_data,
    cmap='cool',
    window=matplotlib.mlab.window_none)
```

The output for the control piano scale in D major audio file is as follows:

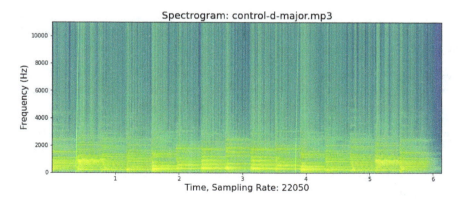

Figure 8.10 – Spectrogram with window_none, piano scale (control-d-major)

The output for the human speech dataset is as follows:

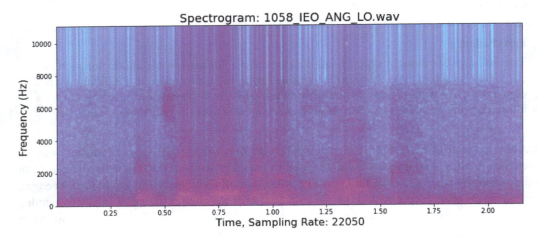

Figure 8.11 – A spectrogram with window_none, voice (1058_IEO_ANG_LO)

The other values for window parameters are `numpy.blackman`, `numpy.bartlett`, `scipy.signal`, and `scipy.signal.get_window`, and the audio from *Figure 8.11* is a woman saying *"It is 11 o'clock."*

Fun challenge

Here is a thought experiment. Given the Spectrogram graph as an image, can you reverse-engineer and play the audio from the picture? A hint is to research the inverse Apectrogram software and theories.

Pluto continues plotting various Spectrograms and color maps because audio engineers may need to exaggerate or highlight a particular frequency or audio property. In addition, the augmentation technique is similar to the previous chapter. Thus, spending more time expanding your insight into the Spectrogram is worthwhile.

Pluto can use parameters individually or combine multiple parameters to produce a different desired outcome, such as using the `sides` parameter on the real-world music dataset and combining `sides` with the `mode` parameters on the control piano scale data. The commands are as follows:

```
# the control piano scale in D major
pluto.draw_spectrogram(pluto.df_music_data,
    cmap='plasma',
    sides='twosided')
# the music dataset
```

```
pluto.draw_spectrogram(pluto.audio_control_dmajor,
    window=matplotlib.mlab.window_none,
    sides='twosided',
    mode='angle')
```

The output for the music with sides equal to twosided is as follows:

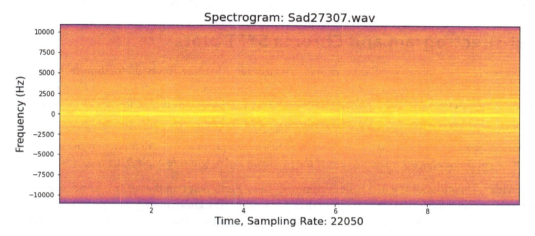

Figure 8.12 – A spectrogram with twosided, music (Sad27307)

The output for the control piano scale audio with sides equal to twosided and mode equal to angle is as follows:

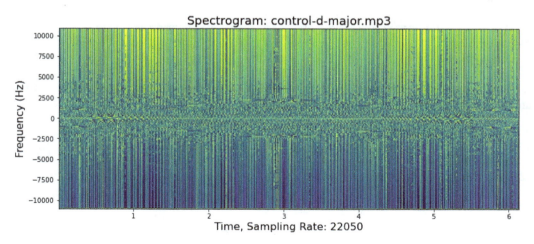

Figure 8.13 – Spectrogram with twosided and angle, music (control-d-major)

> **Fun challenge**
>
> Pluto has additional parameter combinations in the Python Notebook. Thus, it would be best if you modified or hacked the code. It will be fun to experience how different Spectrograms can look for different real-world datasets.

Next are the Mel-spectrogram and Chroma STFT plots. They are similar to a Spectrogram.

Mel-spectrogram and Chroma STFT plots

Pluto spends additional time plotting various Spectrograms because the augmentation technique is the same as in the Waveform graph in *Chapter 7*. Pluto will write fewer new wrapper functions. He will reuse the methods from the previous chapter, but before that, let's draw more Spectrograms.

The subjective unit of pitch, also known as the **mel scale**, is a pitch unit with equal distance between pitches. *S. S. Stevens, John Volkmann, and E. B. Newmann* proposed the mel scale in the scholarly paper titled, *A scale for the measurement of the psychological magnitude of pitch*, in 1937.

The math calculation for the mel scale is complex. Thus, Pluto relies on the melspectrogram() method from the Librosa library to perform the computation. The Pluto draw_melspectrogram() wrapper method uses the Librosa melspectrogram() function, and the code snippet is as follows:

```
# code snippeet for the melspectrogram
mel = librosa.feature.melspectrogram(y=data_amp,
  sr=sam_rate,
  n_mels=128,
  fmax=8000)
mel_db = librosa.power_to_db(mel, ref=numpy.max)
self._draw_melspectrogram(mel_db, sam_rate, data_amp,
  cmap=cmap,
  fname=tname)
```

The entire function code is in the Python Notebook. Pluto draws the Mel-spectrogram for the control piano scale and the human speech datasets are as follows:

```
# Control piano scale
pluto.draw_melspectrogram(pluto.audio_control_dmajor)
# Music dataset
pluto.draw_melspectrogram(pluto.df_voice_data, cmap='cool')
```

The output of the Mel-spectrogram for the control piano scale is as follows:

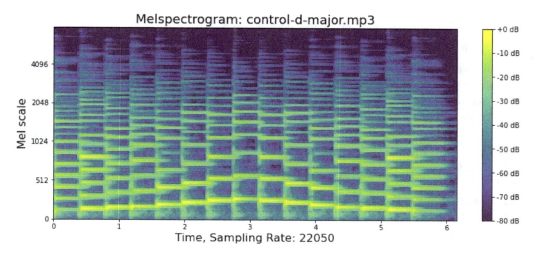

Figure 8.14 – Mel-spectrogram control piano scale (control-d-major)

The output of the Mel-spectrogram for the human speech dataset is as follows:

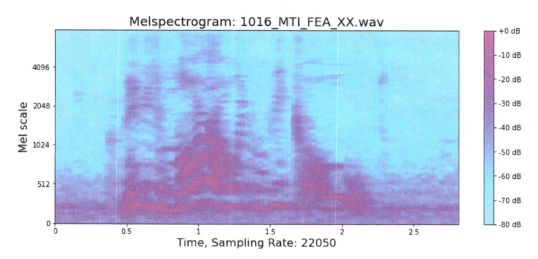

Figure 8.15 – Mel-spectrogram music (1016_MTI_FEA_XX)

Figure 8.15 audio is a man saying *"Maybe tomorrow, it will be cold."* Every Mel-spectrogram has an audio-play button in the Python Notebook, where you can listen to the audio file.

The Chroma STFT is a signal's sinusoidal frequency and local section phase content as it changes over time. *Dr. Dennis Gabor* introduced STFT, also known as the **Gabor transform**, in the scholarly paper, *Theory of Communication*, in 1944 and revised it in 1945.

Chroma STFT is a method of analyzing musical audio signals by decomposing them into their constituent frequencies and amplitudes with respect to time. It is used to characterize the instrument used in a given piece of music and identify unique features in short pieces of music. Chroma STFT is most often used to identify spectral characteristics of a music signal, allowing the components to be quantified and compared to other versions of the same piece.

Pluto adds slightly to the `draw_melspectrogram()` wrapper method to accommodate the Chroma STFT plot. The additional parameter is `is_chroma`, and the default value is `False`. The `_draw_melspectrometer()` helper function does not change. The code snippet is as follows:

```
# code snippet for the chroma_stft
stft = librosa.feature.chroma_stft(data_amp,
  sr=sam_rate)
self._draw_melspectrogram(stft, sam_rate, data_amp,
  cmap=cmap,
  fname=tname,
  y_axis=yax,
  y_label=ylab)
```

The entire function code is on the Python Notebook. Pluto draws the Chroma STFT graphs for the control piano scale, the music, and the urban sound datasets as follows:

```
# Control piano scale
pluto.draw_melspectrogram(pluto.audio_control_dmajor,
  is_chroma=True)
# Music dataset
pluto.draw_melspectrogram(pluto.df_music_data,
  is_chroma=True,
  cmap='plasma')
# Urban sound dataset
pluto.draw_melspectrogram(pluto.df_sound_data,
  is_chroma=True,
  cmap='brg')
```

The output for the Chroma STFT plot for the control piano scale in D major is as follows:

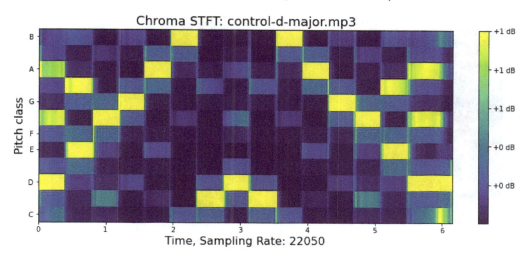

Figure 8.16 – Chroma STFT, control piano scale (control-d-major)

The output for the music dataset is as follows:

Figure 8.17 – Chroma STFT, music (Sad19513)

The output for the urban sound dataset is as follows:

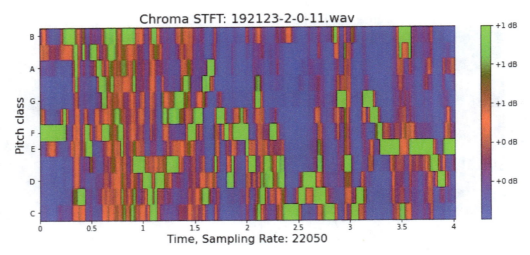

Figure 8.18 – Chroma STFT, urban sound (192123-2-0-11)

Figure 8.17's audio is cinematic music with a strong violin lead, and *Figure 8.18* sounds like noisy kids playing in an outdoor playground.

> **Fun fact**
>
> When generating new images or plots, Pluto automatically writes or exports the image files to the ~/Data-Augmentation-with-Python/pluto_img directory. Thus, Pluto automatically saved the augmented images in *Chapter 3* and *Chapter 4* and the Waveform graph, audio Spectrogram, Mel-spectrogram, and Chroma STFT charts in *Chapter 7* and *Chapter 8*. The helper function name is _drop_image() with the pluto[id].jpg file format, where id is an auto-increment integer from the self.fname_id variable.

We have discussed in detail and written Python code for the audio Spectrogram, Mel-spectrogram, and Chroma STFT. Next, Pluto will describe how to perform audio augmentation with a Spectrogram.

Spectrogram augmentation

Pluto will reuse most of the wrapper functions from *Chapter 7*. You can reread the previous chapter if the following code seems challenging. Pluto will shorten his explanation of the wrapper functions because he assumes you are an expert at writing audio augmentation wrapper functions.

Audio Spectrogram, Mel-spectrogram, Chroma STFT, and Waveform charts take the returned amplitude data and sampling rate from the Librosa load() function reading an audio file. There is an additional transformation of the amplitude data, but they serve the same goal of visualizing the sound wave and frequencies.

After reviewing many scholarly published papers, Pluto concluded that the audio augmentation techniques in *Chapter 7* apply equally well to the audio Spectrogram, Mel-spectrogram, and Chroma STFT. In particular, he referred to the scholarly paper, *Audio Augmentation for Speech Recognition* by Tom Ko, Vijayaditya Peddinti, Daniel Povey, and Sanjeev Khudanpur, published in 2015; *Data augmentation approaches for improving animal audio classification* by Loris Nannia, Gianluca Maguoloa, and Michelangelo Paci, published in 2020; and *Deep Convolutional Neural Networks and Data Augmentation for Environmental Sound Classification* by Justin Salamon and Juan Pablo Bello, published in 2017.

Intuitively, there shouldn't be any difference from the technique in *Chapter 7* because the underlying amplitude data and sampling rate are the same. In other words, you can use *Chapter 7* audio augmentation functions for the audio Spectrogram, Mel-spectrogram, and Chroma STFT, such as the following techniques:

- Time-stretching
- Time-shifting
- Pitch-scaling
- Noise injection
- Polarity inversion
- Low-pass filter
- High-pass filter
- Ban-pass filter
- Low-shelf filter
- High-shelf filter
- Band-stop filter
- Peak filter

There are others, such as `Masking` and `Gaps`. They are available from the `audiomentation` library. The **safe** level mentioned in the previous chapter applies equally to the audio Spectrogram, Mel-spectrogram, and Chroma STFT.

> **Fun fact**
>
> You can alter any Python functions by overriding them in the `correct` class. Pluto functions belong to the `PacktDataAug` class. Thus, you can hack and override any of Pluto's methods by adding the `@add_method(PacktDataAug)` code line before the function definition.

Pluto needs to hack the _audio_transform() helper function and includes the new is_waveform parameter setting the default to True so it will not affect methods in *Chapter 7*. The definition of the new method is as follows:

```
# add is_waveform parameter
@add_method(PacktDataAug)
def _audio_transform(self, df, xtransform,
  Title = '',
  is_waveform = True):
```

The updated code snippet is as follows:

```
# keep the default to be same for Chapter 7, Waveform graph
if (is_waveform):
  # augmented waveform
  self._draw_audio(xaug, sam_rate,
    title + ' Augmented: ' + fname)
  display(IPython.display.Audio(xaug, rate=sam_rate))
  # original waveform
  self._draw_audio(data_amp, sam_rate, 'Original: ' + fname)
# update to use spectrogram, me-spectrogram, and Chroma
else:
  xdata = [xaug, sam_rate, lname, 'Pluto']
  self.draw_spectrogram(xdata)
  self.draw_melspectrogram(xdata)
  self.draw_melspectrogram(xdata, is_chroma=True)
```

Thus, the is_waveform parameter is to use the Waveform graphs in *Chapter 7* or the audio Spectrogram, Mel-spectrogram, and Chroma STFT charts. That's it, and this is why we love coding with Pluto. He follows the best object-oriented coding practices, and his functions are in one class.

Pluto adds the new parameter to the play_aug_time_shift() wrapper function and tests it with the control data. The command is as follows:

```
# augment the audio with time shift
pluto.play_aug_time_shift(pluto.audio_control_dmajor,
  min_fraction=0.8,
  is_waveform=False)
```

The output for the audio Spectrogram is as follows:

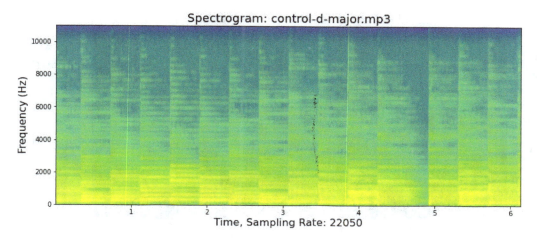

Figure 8.19 – Spectrogram, time shift, piano scale (control-d-major)

The output for the Mel-spectrogram is as follows:

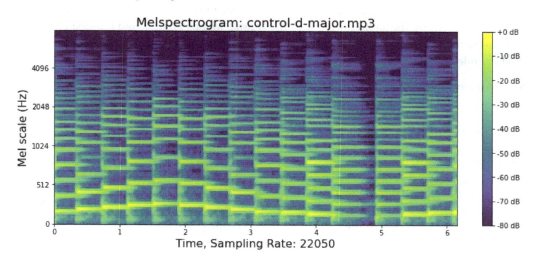

Figure 8.20 – Mel-spectrogram, time shift, piano scale (control-d-major)

The output for the Chroma STFT is as follows:

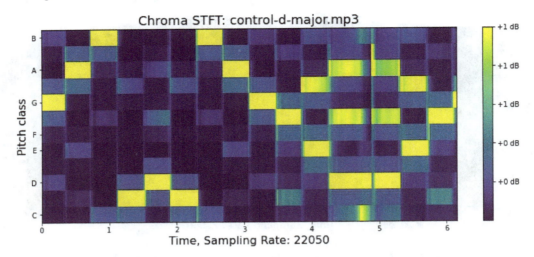

Figure 8.21 – Chroma STFT, time shift, piano scale (control-d-major)

Figure 8.19, *Figure 8.20*, and *Figure 8.21* play the piano scale in D major shift to the left by about 2 seconds. In other words, the audio started with the **G note**, looped around, and finished on the **F# note**. Pluto recommends listening to the before and after effects of the Python Notebook as the easiest method to understand it.

Pluto does the same for the human speech dataset using the following command:

```
# augment audio using time shift
pluto.play_aug_time_shift(pluto.df_voice_data,
    min_fraction=0.8,
    is_waveform=False)
```

The output for the audio Spectrogram is as follows:

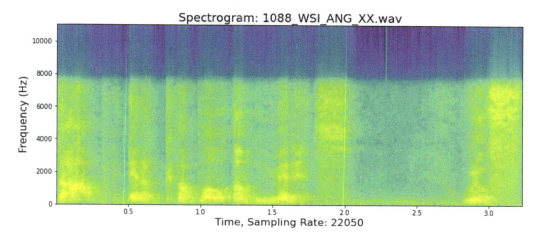

Figure 8.22 – Spectrogram, time shift, human voice (1085_ITS_ANG_XX)

The output for the Mel-spectrogram is as follows:

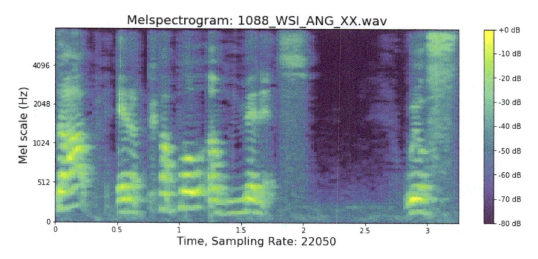

Figure 8.23 – Mel-spectrogram, time shift, human voice (1085_ITS_ANG_XX)

The output for Chroma STFT is as follows:

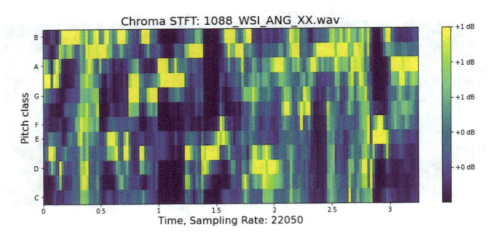

Figure 8.24 – Chroma STFT, time shift, human voice (1085_ITS_ANG_XX)

Figure 8.22, *Figure 8.23*, and *Figure 8.24's* original audio is a man's voice saying "*We will stop in a couple of minutes.*" The augmented version is shifted to "*stop in a couple of minutes [silence] we will.*" Pluto can hear the difference in the before-and-after augmentation effect in the Python Notebook. The goal of audio augmentation is the same for Spectrogram and Waveform graphs, which is to increase the AI accuracy prediction by increasing the input data.

The results for the music and urban sound dataset are shifted similarly. Pluto has the time-shift code in the Python Notebook, where you can run it and see and hear the result. Furthermore, Pluto will skip describing the results for other audio augmentation functions in this chapter. It is because the results are the same as in *Chapter 7*, and the wrapper functions code is in the Python Notebook. However, he will explain the `play_aug_noise_injection()` function because this function can extend to specific topics discussing how sound engineers use Spectrograms.

Sound engineers use standard audio Spectrograms and various other Spectrograms to spot and remove unwanted noises, such as hums, buzz, hiss, clips, gaps, clicks, and pops. Audio augmentation goals are the opposite. We add unwanted noises to the recording within a safe range. Thus, we increase the training datasets and improve the AI prediction accuracy.

Pluto adds white noise to the music dataset using the following command:

```
# augment audio with noise injection
pluto.play_aug_noise_injection(pluto.df_music_data,
  min_amplitude=0.008,
  max_amplitude=0.05,
  is_waveform=False)
```

The output for the audio Spectrogram is as follows:

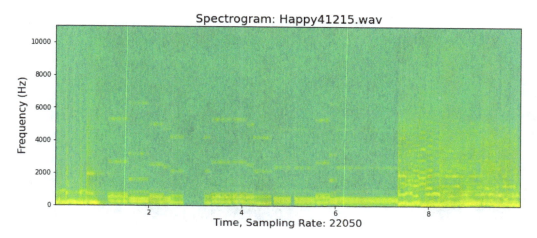

Figure 8.25 – Spectrogram, noise injection, music (Happy41215)

The output for the Mel-spectrogram is as follows:

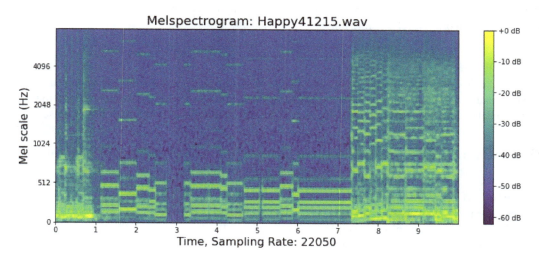

Figure 8.26 – Mel-spectrogram, noise injection, music (Happy41215)

The output for the Chroma STFT is as follows:

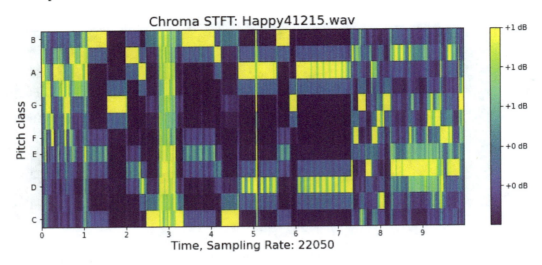

Figure 8.27 – Chroma STFT, noise injection, music (Happy41215)

Figure 8.25, *Figure 8.26*, and *Figure 8.27* play heavy drums, light electronic bells, and heavy electronic guitars with a medium level of white noise.

> **Fun challenge**
>
> Here is a thought experiment. You are part of a team developing a self-driving car system, and your goal is to recognize or identify car honking while driving. How would you augment the audio data? A hint is thinking about real-world driving conditions with traffic or urban noises.

If you have hums, buzz, or pops audio files, you can inject them into the recording by alternating the `play_aug_noise_injection()` wrapper function as follows:

```
# Original use white noise, code snippet
xtransform = audiomentations.AddGaussianNoise(
  min_amplitude=min_amplitude,
  max_amplitude=max_amplitude,
  p=1.0)
# Update to using unwanted noise file
xtransform = audiomentations.AddShortNoises(
  sounds_path="~/path_to_unwanted_noise_file",
  min_snr_in_db=3.0,
  max_snr_in_db=30.0,
  noise_rms="relative_to_whole_input",
  min_time_between_sounds=2.0,
```

```
max_time_between_sounds=8.0,
noise_transform=PolarityInversion(),
p=1.0)
```

The preceding code snippet and complete documentation can be found in the `audiomentations` library on GitHub.

The next topic is a novel idea using a Spectrogram as an image input for deep learning image classification.

Spectrogram images

Fundamentally, audio data is time-series data. Thus AI uses a time-series algorithm, such as the **autoregressive integrated moving average** (**ARIMA**) or **exponential smoothing** (**ES**) algorithm for audio classification. However, there is a better method. You use the Spectrogram as an image representing the audio sound, not the time-series numerical array, for input. Using images as the input data, you can leverage the robust neural network algorithm to classify audio more accurately.

Strictly speaking, this topic does not directly pertain to new audio augmentation techniques. Still, it is an essential topic for data scientists to understand. However, Pluto will not write Python code for building a neural network model using Spectrograms as input.

Deep learning image classification, also known as the machine learning model that uses the artificial neural networks algorithm, achieved an unprecedented accuracy level that exceeds 98% accuracy recently. Many AI scientists apply deep learning techniques to audio datasets, such as *Audio Spectrogram Representations for Processing with Convolutional Neural Networks* by Lonce Wyse, published in 2017, and *Deep Learning Audio Spectrograms Processing to the Early COVID-19 Detection* by Ciro Rodriguez, Daniel Angeles, Renzo Chafloque, Freddy Kaseng, and Bishwajeet Pandey, published in 2020.

The technique takes an audio Spectrogram as the image input, not the audio amplitude, sampling rate, or Mel scale. For example, the music dataset (MEC) goal is to classify a piece of music clip as having a happy or sad mood. Pluto can generate all the audio files to audio Spectrograms and save them on the local drive. He will use the Fast.ai robust AI framework and libraries to create an image classification model. He can achieve 95% accuracy or higher.

The big question is can you use the image augmentation methods discussed in *Chapter 3* and *Chapter 4* to apply to Spectrogram?

It depends on the safe level and the objective of the AI model. For example, using the image augmentation technique, vertically flipping a spectogram involves flipping high to low frequencies and vice versa. Pluto wonders how that would affect the music's mood. It could be an **unsafe** technique. However, image noise injection methods with low noise values could be a safe technique with a Spectrogram. Pluto thinks it is more suitable to stay with the audio augmentation techniques in *Chapter 7*.

Similar deep learning methods can be applied to the human speech (CREMA-D) dataset to classify the age, sex, or ethnicity of the speaker.

> **Fun challenge**
>
> This is a thought experiment. Can you use speech-to-text software to convert the voice into text and use text augmentation functions in *Chapter 5* and *Chapter 6*? A hint is to think about the scope of the project. For example, it could work if the AI aims to infer sentiment analysis but not if the goal is to identify male or female voices.

For the urban sound (US8K) dataset, Pluto could use the deep learning multilabel classification to identify different types of sound in an urban sound clip, such as a jackhammer, wind, kids playing, rain, dogs barking, or gunshots.

> **Fun challenge**
>
> Pluto challenges you to refactor the `Pluto` class to make it faster and more compact. You should also include all the image and text wrapper and helper functions from previous chapters. Pluto encourages you to create and upload your library to *GitHub and PyPI.org*. Furthermore, you don't have to name the class `PacktDataAug`, but it would give Pluto and his human companion a great big smile if you cited or mentioned the book. The code goals were ease of understanding, reusable patterns, and teaching you about the Python Notebook. Thus, refactoring the code as a Python library would be relatively painless and fun.

We have covered audio Spectrogram, Mel-spectrogram, and Chroma STFT representation and augmentation, including the technique of using Spectrograms as image input to the deep learning image classification model. It is time for a summary.

Summary

Audio augmentation is challenging to explain in a book format. Still, we gain a deeper understanding of audio amplitude, frequency, and sampling rate with additional visualization techniques, such as the audio Spectrogram, Mel-spectrogram, and Chroma STFT. Furthermore, in the Python Notebook, you can listen to the before-and-after effects of the audio augmentation.

Compared to the previous chapter, Waveform graphs show the amplitude of a signal over time, giving an understanding of its shape and structure. Spectrogram graphs show a visual representation of the frequencies of a signal over time, providing a deeper insight into the harmonic content of the sound.

An Audio Spectrogram comes in many variations, whether **color mapping, window filtering, spectrum sides, magnitude mode**, or **frequency scale**, among many more in the underlying Matplotlib `specgram()` function. Pluto uses Python code in wrapper functions on a few Spectrogram types. The majority of Spectrogram variations are up to you to explore by expanding the `Pluto` object with additional wrapper functions. Using Pluto's object-oriented best practices, the function wrapper concept, and the audiomentations library, it is easy to expand Pluto with additional wrapper functions.

For Spectrogram augmentation techniques, they are the same techniques as those from *Chapter 7*, such as time -shifting, time-stretching, pitch-scaling, noise injections, bandpass filters, and many others. Intuitively, there should be no difference because in the previous chapter, you choose to visualize the sound wave in Waveform graphs, and in this chapter, you drew them in the audio Spectrogram, Mel-spectrogram, and Chrom STFT plots. Thus, the underlying data is the same.

Pluto has to only modify the `_audio_transform()` helper method with an additional `is_waveform` parameter. The Python code becomes deceptively simple and repetitive afterward, but it hides the robust power of the audiomentations library and Pluto object-oriented best practices.

Throughout the chapter, there were **fun facts** and **fun challenges**. Pluto hopes you will take the advantages provided and expand the experience beyond the scope of this chapter.

The next chapter moves beyond the typical data types, such as image, text, and audio, to tubular data augmentation.

Part 5:
Tabular Data Augmentation

This part includes the following chapter:

- *Chapter 9, Tabular Data Augmentation*

9

Tabular Data Augmentation

Tabular augmentation supplements tabular data with additional information to make it more useful for predictive analytics. Database, spreadsheet, and table data are examples of tabular data. It involves transforming otherwise insufficient datasets into robust inputs for ML. Tabular augmentation can help turn unstructured data into structured data and can also assist in combining multiple data sources into a single dataset. It is an essential step in data pre-processing for increasing AI predictive accuracy.

The idea of tabular augmentation is to include additional information to a given dataset that can then be used to generate valuable insights. These datasets can come from various sources, such as customer feedback, social media posts, and IoT device logs. Tabular augmentation can add new information columns to the dataset by enriching the existing columns with more informative tags. It increases the completeness of the dataset and provides more accurate insights.

Tabular augmentation is an important method to consider when pre-processing and generating insights from data. It provides a way to work with incomplete and unstructured datasets by organizing and enriching them for improved accuracy and speed. By implementing tabular augmentation, you can better unlock the value of real-world datasets and make better-informed decisions.

Tabular augmentation is a young field for data scientists. It is contrary to using analytics for reporting, summarizing, or forecasting. In analytics, altering or adding data to skew the results to a preconceived desired outcome is unethical. In data augmentation, the purpose is to derive new data from an existing dataset. The two goals might be incongruent, but they are not. DL is an entirely different technique from traditional analytics. One is based on a neural network algorithm, while the other is based on statistical analysis and data relationships.

The salient point is that even though you might introduce synthetic data into the datasets, it is an acceptable practice. The *Synthesizing Tabular Data using Generative Adversarial Networks* paper, by Lei Xu and Kalyan Veeramachaneni, published in the *arXiv Forum* in November 2018, supports this proposition.

This chapter focuses on describing concepts. It has a few practical coding examples using the Python Notebook. One main reason for this is that there are only a few tabular augmentation open source libraries available. You will spend most of the coding time plotting various graphs to inspire further insight from the datasets.

Before continuing, let's take a sneak peek at a real-world tabular dataset. Later, Pluto will explain in detail how to write Python code for the following:

```
# print out the tabular data
pluto.df_bank_data[['fraud_bool',
    'proposed_credit_limit',
    'customer_age',
    'payment_type']].sample(5)
```

The output is as follows:

	fraud_bool	proposed_credit_limit	customer_age	payment_type
751785	0	200.0	30	AD
379542	0	200.0	40	AC
208050	0	200.0	40	AD
562807	0	200.0	50	AB
471898	0	1000.0	30	AB

Figure 9.1 – Bank Account Fraud Dataset Suite (NeurIPS 2022)

One challenge in augmenting tabular data is that no fixed methods work universally, such as flipping images, injecting misspelled words, or time-stretching audio files. You will learn that the dataset dictates which augmentation techniques are **safe** or in a **safe range**. It is essential to thoroughly review the tabular dataset before augmenting it.

> **Fun fact**
>
> **Deep neural networks (DNNs)** excel at predicting future stock values and tabular data, based on the scholarly paper *Deep learning networks for stock market analysis and prediction: Methodology, data representations, and case studies*, by Eunsuk Chong, Chulwoo Han, and Frank C. Park. It was published by Elsevier, *Expert Systems with Applications*, Volume 83, on 15 October 2017.

Tabular augmentation is an approach to augmenting a tabular dataset with synthetic data. It involves adding new columns to a tabular dataset with features from the derived calculation. You will spend the majority of the time in Python code visualizing the real-world tabular dataset with exotics plots. In this chapter, we will cover the following topics:

- Tabular augmentation libraries

- Augmentation categories

- Real-world tabular datasets

- Exploring and visualizing tabular data

- Transformation augmentation

- Extraction augmentation

Let's start with augmentation libraries.

Tabular augmentation libraries

Tabular augmentations are not established as image, text, or audio augmentations. Typically, data scientists develop tabular augmentation techniques specific to a project. There are a few open source projects on the GitHub website. Still, DL and generative AI will continue to advance in forecasting for time series and tabular data predictions, and so will tabular augmentations. The following open source libraries can be found on the GitHub website:

- **DeltaPy** is a tabular augmentation for generating and synthesizing data focusing on financial applications such as time series stock forecasting. It fundamentally applies to a broad range of datasets. The GitHub website link is `https://github.com/firmai/deltapy`. The published scholarly paper is called *DeltaPy: A Framework for Tabular Data Augmentation in Python*, by Derek Snow, The Alan Turing Institute, in 2020.

- The **Synthetic Data Vault (SDV)** is for augmenting tabular data by generating synthetic data from a single table, multi-table, and time series data. In 2020, Kalyan Veeramachaneni, Neha Patki, and Saman Amarsinghe developed a commercial version named *Datacebo*. The GitHub link is `https://github.com/sdv-dev/SDV`.

- The tabular **Generative Adversarial Network (GAN)** uses the successfully generating realistic image algorithm and applies it to tabular augmentation. The scholarly paper is *Tabular GANs for uneven distribution*, by Insaf Ashrapov, published by *Cornell University, Arxiv*, in 2020. The GitHub website link is `https://github.com/Diyago/GAN-for-tabular-data`.

Pluto has chosen the **DeltaPy** library to use as the engine under the hood for his tabular augmenting wrapper functions, but first, let's look at the augmentation categories.

Augmentation categories

It is advantageous to group tabular augmentation into categories. The following concepts are new and particular to the DeltaPy library. The augmentation functions are grouped into the following categories:

- **Transforming** techniques can be applied for cross-section and time series data. Transforming techniques in tabular augmentation are used to modify existing rows or columns to create new, synthetic data representative of the original data. These methods can include the following:

  - **Scaling**: Increasing or decreasing a column value to expand the diversity of values in a dataset
  - **Binning**: Combining two or more columns into a single bucket to create new features
  - **Categorical encoding**: Using a numerical representation of categorical data
  - **Smoothing**: Compensating for unusually high or low values in a dataset
  - **Outlier detection and removal**: Detecting and removing points farther from the norm
  - **Correlation-based augmentation**: Adding new features based on correlations between existing features

- The **interacting** function is a cross-sectional or time series tabular augmentation that includes normalizing, discretizing, and autoregression models. In tabular augmentation, these functions are used to specify interactions between two or more variables and help generate new features that represent combinations of the original variables. This type of augmentation is beneficial when modeling the relationships between multiple input features and the target variable, as it allows the model to consider interactions between the different components.

- The **mapping** method, which uses **eigendecomposition** in tabular augmentation, is a method of unsupervised learning that uses data decomposition to transform data into lower-dimensional space using eigenvectors and eigenvalues. This type of feature transformation is useful for clustering, outlier detection, and dimensionality reduction. By projecting the data onto the eigenvectors, the data can be represented in a reduced space while still preserving the structure of the data.

- The **extraction** method is a tabular augmentation technique that utilizes **Natural Language Processing (NLP)** to generate additional information from textual references in tabular datasets. It uses the **TSflesh** library, a collection of rules and heuristics, to extract additional data from text, such as names, dates, and locations. This approach is beneficial in augmenting structured datasets, where the output of **sentence split**, **tokenization**, and **part-of-speech tagging** is used to create features that can be used for further processing.

- **Time series synthesis (TSS)** is a method for tabular data augmentation where rows of data across multiple sources or temporal points in time are synthesized together. You can use it to increase a dataset's size and create a more consistent set of features.

- **Cross-sectional synthesis (CSS)** is a method for tabular data augmentation where columns of data from multiple sources are combined. You can use it to increase a dataset's features and create a more complete and holistic data view.

- The **combining** technique uses the mix-and-match process from the existing methods.

There are functions associated with each category in the DeltaPy library. However, Pluto has to construct a neural network model, such as a **convolutional neural network (CNN)** or **reoccurring neural network (RNN)**, to gauge the effectiveness of these methods. It is a complex process, and Pluto will not implement a CNN in this chapter. Nevertheless, Pluto will demonstrate the mechanics of using the DeltaPy library on the Python Notebook. He will not explain how they work.

Now, it is time to download the real-world datasets from the *Kaggle* website.

Real-world tabular datasets

There are thousands of real-world tabular datasets on the *Kaggle* website. Pluto has chosen two tabular datasets for this process.

The *Bank Account Fraud Dataset Suite (NeurIPS 2022)* contains six synthetic bank account fraud tabular datasets. Each dataset contains 1 million records. They are based on real-world data for fraud detection. Each dataset focuses on a different type of bias. Sergio Jesus, Jose Pombal, and Pedro Saleiro published the dataset in 2022 under the **Attribution-NonCommercial-ShareAlike 4.0 International (CC BY-NC-SA 4.0)** license. The *Kaggle* link is `https://www.kaggle.com/datasets/sgpjesus/bank-account-fraud-dataset-neurips-2022`.

The *World Series Baseball Television Ratings* is a dataset for audiences watching the baseball World Series on television from 1969 to 2022. Matt OP published the dataset in 2022 under the **CC0 1.0 Universal (CC0 1.0) Public Domain Dedication** license. The *Kaggle* link is `https://www.kaggle.com/datasets/mattop/world-series-baseball-television-ratings`.

The steps for instantiating Pluto and downloading real-world datasets from the *Kaggle* website are the same. It starts with loading the `data_augmentation_with_python_chapter_9.ipynb` file into Google Colab or your chosen Jupyter Notebook or JupyterLab environment. From this point onward, the code snippets are from the Python Notebook, which contains the complete functions.

You will be using the code from *Chapter 2* because you will need the wrapper functions for downloading the *Kaggle* dataset, not the wrapper functions for image, text, and audio augmentations. You should review *Chapters 2* and *3* if the steps are unfamiliar:

```
# Clone GitHub repo.
url = 'https://github.com/PacktPublishing/Data-Augmentation-with-
Python'
!git clone {url}
# Initialize Pluto from Chapter 2
```

```
pluto_file = 'Data-Augmentation-with-Python/pluto/pluto_chapter_2.py'
%run {pluto_file}
# Verify Pluto
pluto.say_sys_info()
# Fetch Bank Fraud dataset
url = 'https://www.kaggle.com/datasets/sgpjesus/bank-account-fraud-
dataset-neurips-2022'
pluto.fetch_kaggle_dataset(url)
# Import to Pandas
f = 'kaggle/bank-account-fraud-dataset-neurips-2022/Base.csv'
pluto.df_bank_data = pluto.fetch_df(f)
# Fetch World Series Baseball dataset
url = 'https://www.kaggle.com/datasets/mattop/world-series-baseball-
television-ratings'
pluto.fetch_kaggle_dataset(url)
# Import to Pandas
f = 'kaggle/world-series-baseball-television-ratings/world-series-
ratings.csv'
pluto.df_world_data = pluto.make_dir_dataframe(f)
```

> **Fun challenge**
>
> At the end of *Chapter 8*, Pluto challenged you to refactor the Pluto code for speed and compactness. The goal is to upload Pluto to Pypi.org. This challenge extends that concept and asks you to combine the setup code into one uber wrapper function, such as pluto.just_do_it(). Pluto does not use uber methods because this book aims to make the concepts and functions easier to learn and demystify the process.

The output for gathering Pluto's system information is as follows:

```
-------------------------- : ----------------------------
                              System time : 2023/01/31 07:03
                                 Platform : linux
        Pluto Version (Chapter) : 2.0
                        Python (3.7.10) : actual: 3.8.10 (default,
Nov 14 2022, 12:59:47) [GCC 9.4.0]
                      PyTorch (1.11.0) : actual: 1.13.1+cu116
                        Pandas (1.3.5) : actual: 1.3.5
                           PIL (9.0.0) : actual: 7.1.2
        Matplotlib (3.2.2) : actual: 3.2.2
                              CPU count : 2
                              CPU speed : NOT available
-------------------------- : ----------------------------
```

> **Fun challenge**
>
> Pluto challenges you to search for, download, and import two additional tabular datasets from the *Kaggle* website or your project into pandas.

With that, you have selected a tabular augmentation library, cloned the GitHub repository, instantiated Pluto, and downloaded the two real-world tabular datasets from the *Kaggle* website. Now, it is time for Pluto to explore and visualize the data.

Exploring and visualizing tabular data

Tabular augmentation is more challenging than image, text, and audio augmentation. The primary reason is that you need to build a CNN or RNN model to see the effect of the synthetic data.

Pluto will spend more time explaining his journey to investigate the real-world Bank Fraud and World Series datasets than implementing the tabular augmentation functions using the DeltaPy library. Once you understand the data visualization process, you can apply it to other tabular datasets.

> **Fun fact**
>
> Typically, Pluto starts a chapter by writing code in the Python Notebook for that chapter. It consists of around 150 to 250 combined code and text cells. They are unorganized collections of research notes and try-and-error Python code cells. Once Pluto proves that the concepts and techniques are working correctly through coding, he starts writing the chapter. As part of the writing progress, he cleans and refactors the Python Notebook with wrapper functions and deletes the dead-end code. The clean version of the Python Notebook contains 20% to 30% of the original code and text cells.

In particular, while exploring tabular data, we will cover the following topics:

- Data structure
- First graph view
- Checksum
- Specialized plots
- Exploring the World Series baseball dataset

Let's start with data structures.

Data structure

Pluto starts by inspecting the data structure using pandas' built-in function. He uses the following command:

```
# display tabular data in Pandas
pluto.df_bank_data.info()
```

The result is as follows:

```
<class 'pandas.core.frame.DataFrame'>
RangeIndex: 1000000 entries, 0 to 999999
Data columns (total 32 columns):
 #      Column
Non-Null Count          Dtype
---     ------                                          --------
------          ----
 0      fraud_bool                                      1000000
non-null     int64
 1      income
1000000 non-null     float64
 2      name_email_similarity                   1000000
non-null     float64
 3      prev_address_months_count        1000000 non-null     int64
 4      current_address_months_count     1000000 non-null     int64
 5      customer_age                                    1000000
non-null     int64
 6      days_since_request                        1000000
non-null     float64
 7      intended_balcon_amount              1000000
non-null     float64
 8      payment_type                                    1000000
non-null     object
 9      zip_count_4w                                    1000000
non-null     int64
 10     velocity_6h                                     1000000
non-null     float64
 11     velocity_24h                                    1000000
non-null     float64
 12     velocity_4w                                     1000000
non-null     float64
 13     bank_branch_count_8w              1000000
non-null     int64
 14     date_of_birth_distinct_emails_4 1000000 non-null     int64
 15     employment_status                          1000000
non-null     object
 16     credit_risk_score                          1000000
non-null     int64
 17     email_is_free                              1000000
non-null     int64
```

```
18      housing_status                          1000000
non-null     object
19      phone_home_valid                        1000000 non-null
     int64
20      phone_mobile_valid                      1000000 non-null
   int64
21      bank_months_count                       1000000 non-null
   int64
22      has_other_cards                         1000000 non-null
   int64
23      proposed_credit_limit           1000000 non-null
float64
24      foreign_request                         1000000 non-null
   int64
25      source
1000000 non-null     object
26      session_length_in_minutes       1000000 non-null
float64
27      device_os                                  1000000
non-null     object
28      keep_alive_session                      1000000 non-null
   int64
29      device_distinct_emails_8w       1000000 non-null     int64
30      device_fraud_count                      1000000 non-null
   int64
31      month
1000000 non-null     int64
dtypes: float64(9), int64(18), object(5)
memory usage: 244.1+ MB
```

The Bank Fraud dataset consists of 32 columns, 1 million records or rows, no null values, and five columns that are not numeric. Pluto wants to find out which columns are **continuous** or **categorical**. He does this by calculating the unique value in each column. He uses the following pandas function:

```
# count uniqueness
pluto.df_bank_data.nunique()
```

The partial output is as follows:

```
fraud_bool
  2
income
    9
name_email_similarity                           998861
prev_address_months_count                       374
current_address_months_count            423
customer_age                                         9
```

The Python Notebook contains the complete result. There are 7 continuous columns and 25 categorical columns. Generally, continuous columns have many unique values, as in total records, while categorical columns have unique values between two and a few hundred.

Before using plots to display the data, Pluto will view sample data from the Bank Fraud dataset with the following command:

```
# display the tabular data using Pandas
pluto.df_bank_data[['fraud_bool',
    'proposed_credit_limit',
    'customer_age',
    'payment_type']].sample(5)
```

The output is as follows:

	fraud_bool	proposed_credit_limit	customer_age	payment_type
331487	0	200.0	50	AB
877247	0	200.0	30	AB
369957	0	1500.0	20	AB
934633	0	500.0	40	AD
751676	0	200.0	40	AD

Figure 9.2 – Sample Bank Fraud data

After repeatedly running the command and variation, Pluto finds no surprises in the data. It is clean. The Python Notebook contains additional inspecting functions, such as the pandas `describe()` method.

> **Fun fact**
>
> For a tabular dataset, you will write custom code for inspecting, visualizing, and augmenting the data. In other words, there will be more reusable concepts and processes than reusable code being carried over to the next project.

The Bank Fraud dataset has 32 million elements, which is the typical size of data that data scientists work with. However, your Python Notebook would crash if you tried to plot 32 million points using pandas and Matplotlib with the default settings. Pluto created a simple graph, `pluto.df_bank_data. plot()`, and his Google Colab Pro-version Python Notebook crashed. It required additional RAM.

First graph view

The various plots are not directly aiding in the tabular augmentation process. The primary goal is for you to envision a sizeable tabular dataset. Reading millions of data points is less effective than seeing them plotted on a graph. You may skip the sections about plotting and go directly to the tabular augmentation techniques using the DeltaPy library.

For a large dataset, the solution is to select graphs with calculated or summarizing values. Hence, there will be fewer points to plot. For example, the **histogram** graph is a viable choice because it groups the frequency of ranges. Pluto uses a wrapper function to draw the histogram plot:

```
# display histogram plot
pluto.draw_tabular_histogram(pluto.df_bank_data,
    title='Bank Fraud data with 32 million points')
```

The key code line for the wrapper function is as follows:

```
# code snippet, use Pandas histogram function
df.plot.hist()
```

The output is as follows:

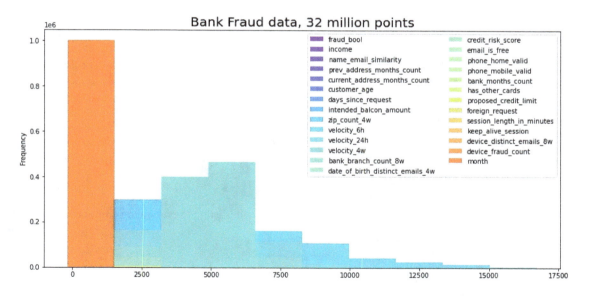

Figure 9.3 – Bank Fraud histogram plot

Figure 9.3 does not yield any beneficial insights. Thus, Pluto proceeds to summarize the data with a **checksum** concept.

Checksum

Pluto spends weeks playing with different types of graphs and graphing packages such as **Matplotlib**, **Seaborn**, **Joypi**, and **PyWaffle**. He has fun, but most do not enhance the visualization of the Bank Fraud and World Series datasets.

At this point, Pluto will get back to more plotting. In tabular data, displaying the string, non-numeric data is challenging. A clean solution is transforming the categorical string data into an integer token index. Pluto writes the `_fetch_token_index()` helper function to index value from a list. The key code snippet is as follows:

```
# code snippet for token index
for i, x in enumerate(xarr):
  if (val == x):
    return i
```

The `add_token_index()` wrapped function uses the helper function and the pandas `apply()` function. The essential code snippet is as follows:

```
# code snippet for tokenize
arrname = numpy.array(df[cname].unique())
df[tname] = df[cname].apply(
  self._fetch_token_index,
  args=(arrname,))
```

To put it all together, Pluto uses the following command to copy and create the tokenized columns for the Data Fraud dataset:

```
# tokenize the data
pluto.df_bank_tokenize_data = pluto.df_bank_data.copy()
pluto.add_token_index(
  pluto.df_bank_tokenize_data,
  ['payment_type', 'employment_status',
  'housing_status', 'source', 'device_os'])
```

Pluto double-checked the tokenization by viewing sample values using the following commands:

```
# print out first 6 row of the tabular data
pluto.df_bank_tokenize_data[['payment_type',
  'payment_type_tokenize']].head(6)
```

The output is as follows:

	payment_type	payment_type_tokenize
0	AA	0
1	AB	1
2	AC	2
3	AB	1
4	AB	1
5	AD	3

Figure 9.4 – Bank Fraud sample tokenized data

Pluto double-checked the other columns, and they are correct. You can view the code and the results by reading the Python Notebook.

For data analysis, it is practical to have a **checksum** column where a number represents each row. It could be a summation, average, or a complex formula of the elements' relationship with weighted value. Pluto's _fetch_checksum() helper function uses the pandas apply() method with lambda. The code snippet is as follows:

```
# code snippet for calculate the checksum
df['checksum'] = df.apply(
  lambda x: numpy.mean(tuple(x)), axis=1)
```

Pluto calculates the checksum for the Bank Fraud dataset using the following command:

```
# compute the checksum
pluto._fetch_checksum(pluto.df_bank_tokenize_data)
```

It took 27 seconds to compute the checksum for 32 million data points. Now, let's explore a few specialized plots with the **checksum** concept.

Specialized plots

Pluto wants to remind you that the following graphs and exercises do not directly pertain to tabular augmentation. The goal is to sharpen your skills in understanding and visualizing sizeable real-world datasets – for example, the Bank Fraud dataset consists of 1 million records in preparation for data augmentation. You can skip the plotting exercises and jump directly to the tabular augmentation lessons if you wish.

Pluto creates `self.df_bank_half_data` with a limited number of columns for ease of display. He uses **Seaborn's** `heatmap()` function to draw the **correlogram** plot. The command is as follows:

```
# plot correlogram
pluto.draw_tabular_correlogram(pluto.df_bank_half_data,
  title='Bank Fraud half Correlogram')
```

The output is as follows:

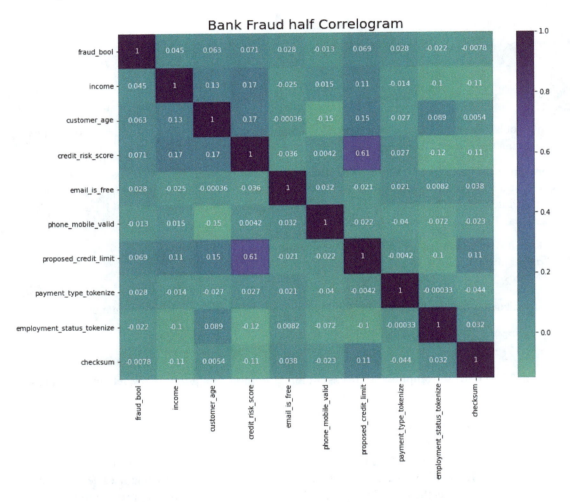

Figure 9.5 – Bank Fraud half correlogram

Figure 9.5 shows a high relationship between `credit_risk_score` and `proposed_credit_limit` with 61%. `fraud_bool` has a low correlation with all other parameters.

When Pluto draws the correlogram plot with the entire dataset, it exposes a high correlation between the **checksum** and **velocity_6h**, **velocity_24h**, and **velocity_4w**. The code and the output can be found in the Python Notebook.

The `draw_tabular_heatmap()` wrapper function looks like a heatmap. The command is as follows:

```
# plotting heatmap
pluto.draw_tabular_heatmap(
    pluto.df_bank_tokenize_data,
    x='checksum',
    y='month')
```

The output is as follows:

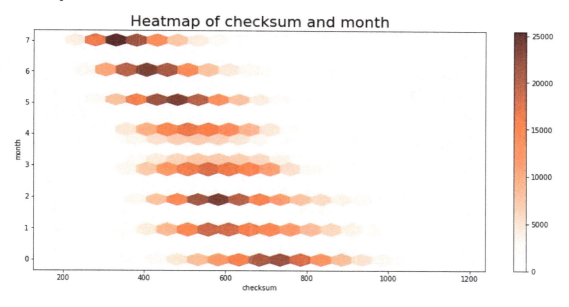

Figure 9.6 – Bank Fraud checksum and month heatmap

Figure 9.6 shows a pattern, but the relationship between the **checksum** and **month** is unclear.

Fun fact

Pluto is not an expert in reading Bank Fraud data, and it is natural for you not to be an expert in every domain. Pluto consults friends in banking and consumer protection agencies for background research. Here are a few charts that he uses in his work.

The fraud data, `fraud_bool == 1`, is 1% of the total. Thus, Pluto might want to augment more fraud data. He creates a pandas DataFrame using the following commands:

```
# tokenize the text or categorical columns
pluto.df_bank_fraud_data = pluto.df_bank_tokenize_data[
  pluto.df_bank_tokenize_data.fraud_bool == 1]
pluto.df_bank_fraud_data.reset_index(
  drop=True,
  inplace=True)
```

The following two graphs suggested by Pluto's banking expert friends are fun to create but may not benefit the Bank Fraud augmentation. The complete code is in the Python Notebook. Nevertheless, they are thought-provoking concepts over the standard line or bar charts:

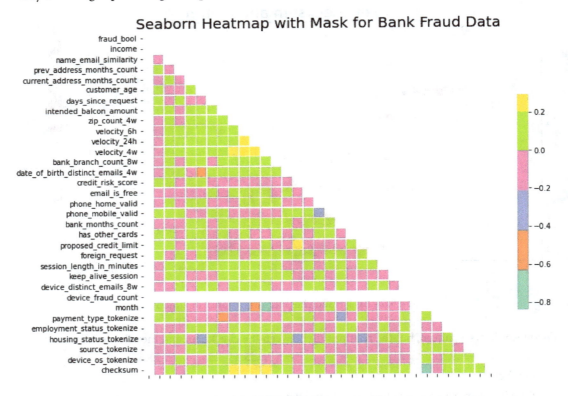

Figure 9.7 – Bank Fraud Seaborn heatmap with mask

The next graph is the Swarmplot.

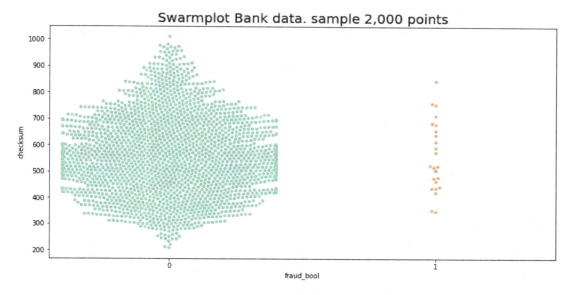

Figure 9.8 – Bank Fraud Seaborn swarm plot

Fun challenge

Can you make use of the `tripcolor()` 3D graph, shown in *Figure 9.9*, using the Bank Fraud dataset? The `tripcolor()` code is in the Python Notebook:

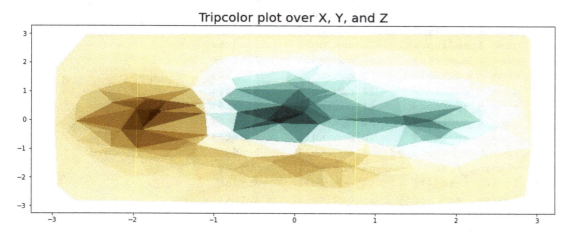

Figure 9.9 – Fun challenge – a tripcolor plot of random values

Exploring the World Series data

In this section, Pluto will spend much time plotting various graphs to understand and visualize the World Series data. He is not performing tabular augmentation. Even though comprehending the data is essential before deciding which augmentation functions are applicable, you can skip this exercise and directly go to the tabular augmentation wrapper functions.

> **Fun fact**
>
> Anecdotally, Pluto, an imaginary Sybirian Huskey, loves to rush ahead and start writing augmenting code without taking the time to sniff out the content of the datasets. Consequently, his AI model diverged 9 out of 10 times, resulting in high levels of false negatives and false positives. Thus, spending 40% to 70% of the time studying the datasets seems non-productive, but it is not. It is an acceptable fact when working with real-world datasets.

Pluto follows a similar process for the World Series dataset. He runs the first `info()` method, followed by `nunique()`, `describe()`, and then `sample()`. The World Series dataset consists of 14 columns and 54 rows, representing 756 data points. There are 11 numeric columns and three label categories. Other factors are eight **continuous** and six **categorical** columns. The output of the `pluto.df_world_data.info()` command is as follows:

```
# describe the tabular dataset
<class 'pandas.core.frame.DataFrame'>
RangeIndex: 54 entries, 0 to 53
Data columns (total 14 columns):
 #   Column              Non-Null Count   Dtype
---  ------              --------------   -----
 0   year                54 non-null      int64
 1   network             54 non-null      object
 2   average_audience    54 non-null      int64
 3   game_1_audience     53 non-null      float64
 4   game_2_audience     52 non-null      float64
 5   game_3_audience     53 non-null      float64
 6   game_4_audience     53 non-null      float64
 7   game_5_audience     44 non-null      float64
 8   game_6_audience     31 non-null      float64
 9   game_7_audience     18 non-null      float64
 10  total_games_played  54 non-null      int64
 11  winning_team        54 non-null      object
 12  losing_team         54 non-null      object
 13  losing_team_wins    54 non-null      int64
dtypes: float64(7), int64(4), object(3)
memory usage: 6.0+ KB
```

Other results can be found in the Python Notebook. The histogram plot is the practical first data visualization technique for the World Series dataset. The command is as follows:

```
# plot histogram graph
pluto.draw_tabular_histogram(pluto.df_world_data,
  title='World Series Baseball',
  maxcolors=14)
```

The output is as follows:

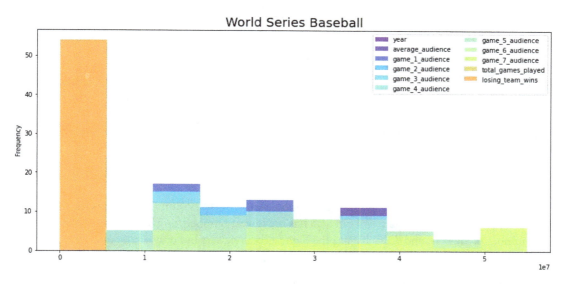

Figure 9.10 – World Series histogram

The histogram plot shown in *Figure 9.10* does not highlight the comparison between the audience in the seven games. Pluto uses the `joyplot()` method from the **joypy** library to display the relationship between the viewing audience and the TV networks. The command is as follows:

```
# plot joyplot graph
pluto.draw_tabular_joyplot(pluto.df_world_data,
  x=['game_1_audience', 'game_2_audience', 'game_3_audience',
     'game_4_audience', 'game_5_audience', 'game_6_audience',
     'game_7_audience'],
  y='network',
  t='World series baseball audience')
```

The output is as follows:

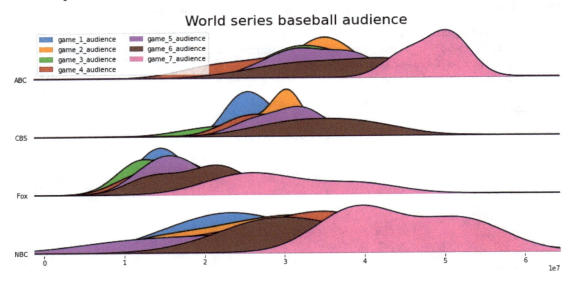

Figure 9.11 – World Series audience and TV networks

Figure 9.11 is a beautiful and insightful visualization plot. NBC television network has the highest number of game viewers for game #7 but also has the lowest number for game #5. Fox TV has the least number of viewers, and ABC TV has the highest total viewers but only a little more than NBC TV. Game #3 has the lowest number of viewers, while game #7 has the highest.

Pluto prepares the World Series dataset for augmenting by converting the label categories into integer token indexes and calculates the checksum. The commands are as follows:

```
# copy tokenize data
pluto.df_world_tokenize_data = pluto.df_world_data.copy()
# eliminate the null value
pluto.df_world_tokenize_data=pluto.df_world_tokenize_data.fillna(0)
# tokenize the data
pluto.add_token_index(pluto.df_world_tokenize_data,
  ['network', 'winning_team', 'losing_team'])
pluto.df_world_tokenize_data =
  pluto.df_world_tokenize_data.drop(
  ['network', 'winning_team', 'losing_team'],
  axis=1)
# calculate the checksum
pluto._fetch_checksum(pluto.df_world_tokenize_data)
```

The code for double-checking and printing the results for the tokenization and checksum can be found in the Python Notebook. Pluto made a quick correlogram plot with the following command:

```
# draw the correlogram graph
pluto.draw_tabular_correlogram(pluto.df_world_tokenize_data,
    title='World Series Baseball Correlogram')
```

The result is as follows:

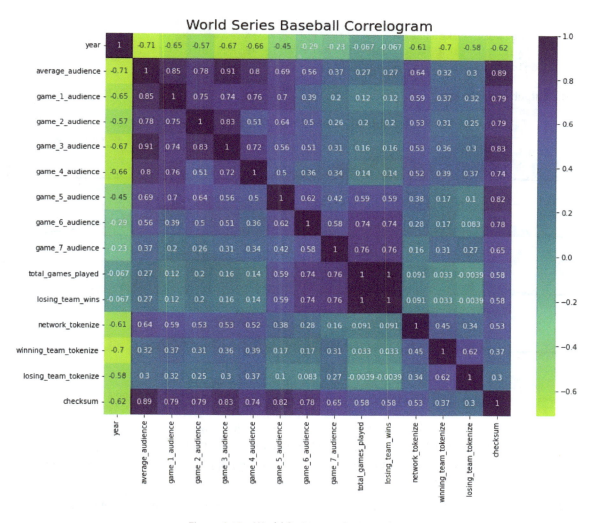

Figure 9.12 – World Series correlogram plot

Figure 9.12 exposes many intriguing relationships between the data. For example, there is a 100% correlation between **losing_team_wins** and **total_game_played**, and strong relationships between **average_audience, game_1_audience, game_2_audience, game_3_audience**, and **game_4_audience**.

Pluto uses the `joyplot()` method to compare the checksum with the average viewers grouped by the TV networks. The command is as follows:

```
# draw the joyplot graph
pluto.draw_tabular_joyplot(pluto.df_world_tokenize_data,
  x=['checksum', 'average_audience'],
  y='network_tokenize',
  t='World series baseball, checksum and average auidence',
  legloc='upper right')
```

The output is as follows:

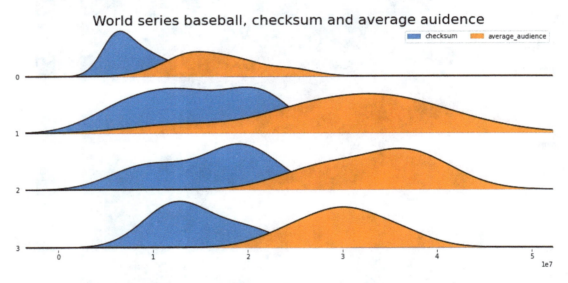

Figure 9.13 – World Series checksum, average audience grouped by TV networks

In *Figure 9.13*, Pluto uses the `mean()` function to calculate the checksum values. Thus, the comparison to the average viewers yields a similar shape. Compared to *Figure 9.11*, the relationship between average audience size and each game's total is not immediately apparent because CBS TV has the highest average but seems to have lower per-game viewers.

At this stage, Pluto wonders if plotting more graphs would help him understand the dataset better. There is a good chance that you are thinking the same thoughts.

The justification for exploring additional charts is twofold. The real-world tabular data is diverse. Thus, knowing various graphs makes you better prepared to tackle your next project. Second, no criteria or algorithm lets you know you have learned about the datasets sufficiently. Therefore, if you know the data, skip to the tabular augmentation functions section or follow along with Pluto as he learns new graphs.

Pluto uses the **waffle** graph to visualize the winning and losing team count. The wrapper function `draw_tabular_waffle()` uses the `Waffle` class from the **pywaffle** library. The command for displaying the World Series winning teams is as follows:

```
# plot the waffle graph
pluto.draw_tabular_waffle(pluto.df_world_data,
  col='winning_team',
  title='World Series Baseball Winning Team')
```

The output is as follows:

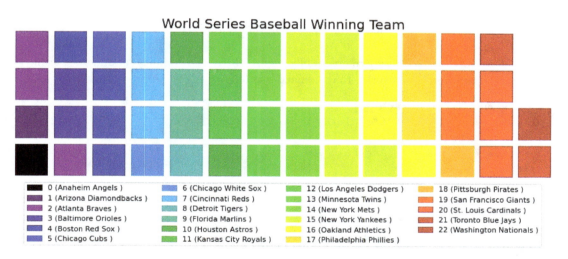

Figure 9.14 – World Series winning team

Pluto does the same for displaying the losing teams. The command is as follows:

```
# draw the waffle graph
pluto.draw_tabular_waffle(pluto.df_world_data,
  col='losing_team',
  title='World Series Baseball Losing Team')
```

The output is as follows:

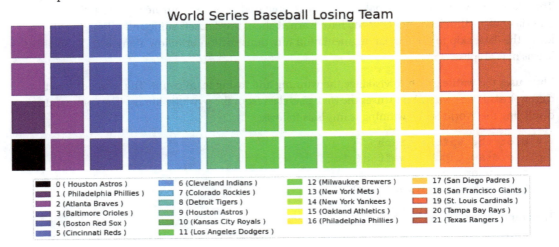

Figure 9.15 – World Series losing team

Figures 9.14 and *9.15* are beautifully colored waffle graphs. There is no dominant or underdog team. Pluto does the same for TV networks. The command is as follows:

```
# draw the waffle graph
pluto.draw_tabular_waffle(pluto.df_world_data,
  col='network',
  title='World Series Baseball Network',
  anchor=(0.5, -0.2))
```

The output is as follows:

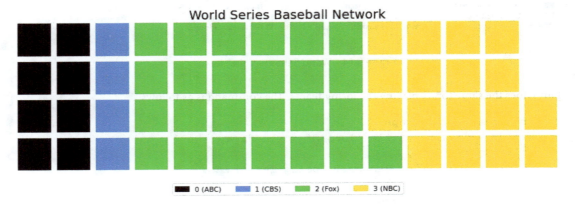

Figure 9.16 – World Series TV networks

Figure 9.16 yields a surprising hidden fact in the data: Fox TV aired the most games, but from *Figures 9.11* and *9.12*, it does not seem like the network with the most viewers.

> **Fun challenge**
>
> Here is a thought experiment: can you visualize a **four-dimensional** (**4D**) graph? Hint: a 2D chart displays two measurements, such as the number of TV audiences per game, or one vector with an implied time series as the X-axis, such as the bank member's income with the X-axis indicated as day or month. A 3D graph typically reveals the snow depth level on a mountain. Time could be the fourth dimension.

Pluto has explored and explained the real-world Bank Fraud and World Series datasets. He uses pandas functions to display statistical information and provides numerous graphs to visualize them. Understanding and visualizing the data is the first and most essential step before augmenting tabular data.

> **Fun fact**
>
> Data augmentation is a secret for DL and generative AI to achieve unprecedented accuracy and success. Many scholarly papers reinforced data augmentation's significance, such as *Enhancing Performance of Deep Learning Models with Different Data Augmentation Techniques: A Survey*, by Cherry Khosla and Baljit Singh Saini, published by *IEEE 2020 Intelligent Engineering and Management (ICIEM), International Conference*.

Transforming augmentation

Before digging into the tabular augmentation methods, Pluto will reiterate that he will not build a neural network model to test if the augmentation benefits the particular dataset. In addition, the pattern for writing the wrapper functions follows the previous practice: using the chosen library to do the critical augmentation step.

As the Python Notebook notes, the DeltaPy library's dependency is the **fbprofet** and **pystan** libraries. The three libraries are in beta and may be unstable. Pluto has repeatedly tested the Python code. Once the libraries have been loaded, the code works flawlessly.

Tabular transformation is a collection of techniques that take one variable and generate a new dataset based on the transformation method. It applies to both cross-section and time series data. The DeltaPy library defines 14 functions for transformation.

These transformation techniques include the **operations** functions used in present information, the **smoothing** method used with past data, and the **select filters** procedure used with lagging and leading values.

In image augmentation, Pluto can run the functions and see what changes in the photo. Here, the effects are apparent, such as **cropped**, **enlarged**, or **altered** hue values. Tabular augmentation requires knowledge of DL and time series data. In other words, the output effects are not obvious; therefore, selecting augmentation functions for a particular dataset can be intimidating. Pluto will demonstrate how to write Python code for tabular augmentation, but he will not thoroughly explain when to use them.

Time series forecasting is a mature and highly researched branch of AI. It could take several college courses to understand a time series and how to forecast or predict future outcomes. A compact definition of a time series is a data sequence that depends on time. Typical time series data is the market stock value. For example, Pluto uses the Microsoft stock data for the previous 10 years to predict the closing price of tomorrow, next week, or next month. Weather forecasting is another widespread use of time series algorithms.

The two key concepts in time series data are **lag time** and **windows**. The lag time is from the observer to a set point, while the window is the range of elements segmented. There are dozens of other key concepts in time series algorithms, from the earliest **long short-term memory** (LSTM) neural network to **ARIMA, SARIMA, HWES, ResNet, InceptionTime, MiniRocket**, and many others.

Most tabular data can be converted into time series data. The **World Series** data is a time series based on the year. The **Bank Fraud** data does not directly have a time vector. However, by adding time data, Pluto can predict at which hour of the day, be it early morning or late night, when most online bank fraud occurs, or he can forecast when most bank fraud happens seasonally, such as around Christmas or college Spring Break.

There are 14 transformation methods, and in particular, Pluto will cover the following three functions:

- Robust scaler

- Standard scaler

- Capping

Let's start with the robust scaler.

Robust scaler

The **K-means** and **principal component analysis** (PCA) time series algorithms use Euclidean distance. Thus, scaling applies to the World Series dataset. When you're unsure of the data distribution, the robust scaler, also known as **normalization**, is a viable technique. The algorithm forecasts future outcomes.

Pluto's `augment_tabular_robust_scaler()` wrapped function uses the DeltaPy library function and the joy and waffle plots. The essential code snippet is as follows:

```
# define robust scaler
def augment_tabular_robust_scaler(self, df):
  return deltapy.transform.robust_scaler(df.copy(),
    drop=["checksum"])
```

The full function code can be found in the Python Notebook. The command for the World Series data is as follows:

```
# augment using robust scaler
df_out = pluto.augment_tabular_robust_scaler(
  pluto.df_world_tokenize_data)
# plot joy plot
pluto.draw_tabular_joyplot(df_out,
  x=['game_1_audience', 'game_2_audience', 'game_3_audience',
     'game_4_audience', 'game_5_audience', 'game_6_audience',
     'game_7_audience'],
  y='network_tokenize',
  t='World series baseball audience')
```

The output is as follows:

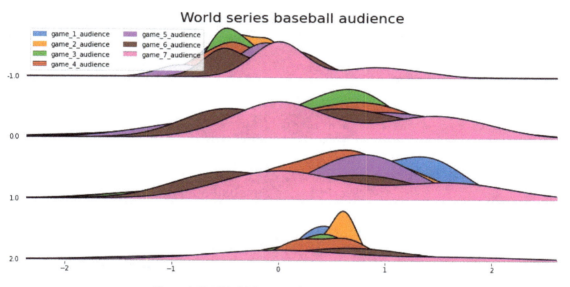

Figure 9.17 – World Series and robust scaler joy plot

Figure 9.17 confirmed that Pluto successfully implemented the robust scaler augmenting technique. Whether it is practical in forecasting is another question entirely. It depends on the goal of the prediction and the base DL model or algorithm used.

The **standard scaler** is similar to the robust scaler.

Standard scaler

DL models that rely on Gaussian distributions or linear and logistic regressions will benefit from the standardization scaler augmentation method. Pluto's `augment_tabular_standard_scaler()` wrapper function uses the DeltaPy library function and the joy and waffle plots. The essential code snippet is as follows:

```
# define standard scaler
def augment_tabular_standard_scaler(self, df):
  return deltapy.transform.standard_scaler(df.copy(),
    drop=["checksum"])
```

The full function code can be found in the Python Notebook. The command is as follows:

```
# augment using standard scaler
df_out = pluto.augment_tabular_standard_scaler(
  pluto.df_world_tokenize_data)
# draw using joy plot
pluto.draw_tabular_joyplot(df_out,
  x=['game_1_audience', 'game_2_audience', 'game_3_audience',
    'game_4_audience', 'game_5_audience', 'game_6_audience',
    'game_7_audience'],
  y='network_tokenize',
  t='World series baseball audience',
  legloc='upper right')
```

The output is as follows:

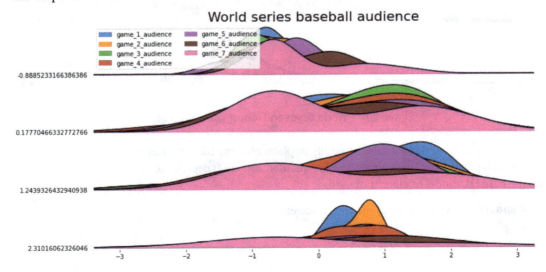

Figure 9.18 – World Series and standard scaler joy plot

Figure 9.18 demonstrated that Pluto did the augmentation correctly. He did not build and train a DL model using the augmented data to confirm that it increased the forecast accuracy. Many tabular augmentation methods require defining the goal for the DL project to verify if the augmentation is beneficial. For example, Pluto could build a DL model for predicting the audience size for the next World Series.

The next tabular transformation technique we'll look at is **capping**.

Capping

The capping technique limits the distribution value, such as average, maximum, minimum, or arbitrary values. In particular, it restricts the values using statistical analysis and replaces the outliers with specific percentile values.

Pluto's `augment_tabular_capping()` wrapper function uses the DeltaPy library function and correlogram plots. The essential code snippet is as follows:

```
# define capping
def augment_tabular_capping(self, df):
  x, y = deltapy.transform.outlier_detect(df, "checksum")
  return deltapy.transform.windsorization(df.copy(),
    "checksum",
    y,
    strategy='both')
```

The command for the Bank Fraud data is as follows:

```
# augment using capping
df_out = pluto.augment_tabular_capping(
  pluto.df_bank_tokenize_data)
# draw correlogram plot
pluto.draw_tabular_correlogram(df_out,
  title='Bank Fraud Capping Transformation')
```

The output is as follows:

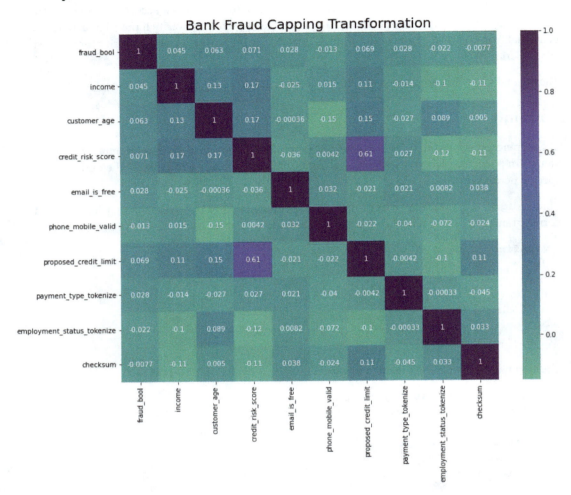

Figure 9.19 – Bank Fraud capping correlogram plot, half data

Figure 9.19 indicates that Pluto implemented the capping technique correctly. Compared to *Figure 9.4*, the original data, the values are similar, as expected.

The Python implementation of tabular transformation wrapper functions becomes repetitive. Thus, Pluto will provide a brief explanation of the other nine methods in the DeltaPy library. They are as follows:

- **Operations** is a technique for using power, log, or square root functions to replace elements in the dataset

- **Smoothing** is a technique that uses the triple exponential smoothing or Holt-Winters exponential smoothing function

- **Decomposing** is a technique that uses the naive decomposition function for seasonal vectors in time series data

- **Filtering** is a technique that uses the Baxter-King bandpass filter to smooth time series data

- **Spectral analysis** is a technique that uses the periodogram function to estimate the spectral density

- **Waveforms** is a technique that uses the continuous harmonic wave radar function to augment waveform data

- **Rolling** is a technique that uses mean or standard deviation to calculate the rolling average over a fixed window size in time series data

- **Lagging** is a technique that calculates the lagged values in time series data

- **Forecast model** is a technique that uses the prophet algorithm to forecast seasonal trends, such as weekly or yearly, in time series data

> **Fun challenge**
> Pluto challenges you to implement three wrapper functions in the Python Notebook from the nine tabular transformation techniques mentioned.

Now that we've reviewed various tabular transformation techniques, let's look at **interaction** techniques.

Interaction augmentation

Interaction techniques are used in ML and statistical modeling to capture the relationships between two or more features in a dataset for augmentation. The goal is to create new augmentation data that captures the interaction between existing components, which can help improve model performance and provide additional insights into the data. You can apply these techniques to cross-sectional or time-specific data, including normalizing, discretizing, and autoregression models.

Pluto has selected two out of seven methods for a hands-on Python programming demonstration. As with the transformation augmentation methods, the coding is repetitive. Thus, Pluto will provide fun challenges for the other five interaction augmentation techniques.

Pluto will start with the **regression** method, then the **operator** method.

Regression augmentation

The regression method uses the **lowess smoother** function to smooth the curve of the data by locally weighting the observations near a given point. It is a useful tabular augmentation technique for exploring relationships in scatterplots where the relationship between the dependent and independent variables needs to be well-described by a linear function. This method can suffer from forward-looking bias. Thus, Pluto recommends caution in using it for predictive modeling.

Pluto's `augment_tabular_regression()` wrapper function uses the DeltaPy library function, a joy plot, and a correlogram graph. The essential code snippet is as follows:

```
# define regression
def augment_tabular_regression(self, df):
  return deltapy.interact.lowess(
    df.copy(),
    ["winning_team_tokenize","losing_team_tokenize"],
    pluto.df_world_tokenize_data["checksum"],
    f=0.25, iter=3)
```

The command for the World Series data is as follows:

```
# augment using tabular regression
df_out = pluto.augment_tabular_regression(
  pluto.df_world_tokenize_data)
# draw joy plot
pluto.draw_tabular_joyplot(df_out,
  x=['game_1_audience', 'game_2_audience', 'game_3_audience',
     'game_4_audience', 'game_5_audience', 'game_6_audience',
     'game_7_audience'],
  y='network_tokenize',
  t='World series baseball audience: Regression',
  legloc='upper right')
```

The output is as follows:

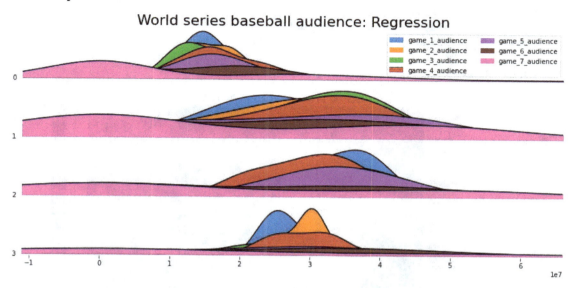

Figure 9.20 – World Series regression augmentation, joy plot

Figures 9.20 confirm that Pluto implemented the regression tabular augmentation correctly. The DeltaPy library does the actual calculation. Thus, if Pluto made a mistake, the result would be an error, or the dataset would contain random numbers and not display correctly. Pluto can only claim the effectiveness of the regression technique to the World Series data. The next tabular augmentation technique we'll look at is the operator augmenting method.

Operator augmentation

The operator method is a simple multiplication or division function between two variables in tabular data.

Pluto's `augment_tabular_operator()` wrapper function uses the DeltaPy library function and a correlogram graph. The essential code snippet is as follows:

```
# define tabular operator
def augment_tabular_operator(self, df):
  return deltapy.interact.muldiv(
    df.copy(),
    ["credit_risk_score","proposed_credit_limit"])
```

Pluto runs the command for the Bank Fraud data, as follows:

```
# augment using tabular operator
df_out = pluto.augment_tabular_operator(
```

```
    pluto.df_bank_tokenize_data)
# draw the correlogram plot
pluto.draw_tabular_correlogram(df_out,
    title='Bank Fraud Operator Interaction')
```

The output is as follows:

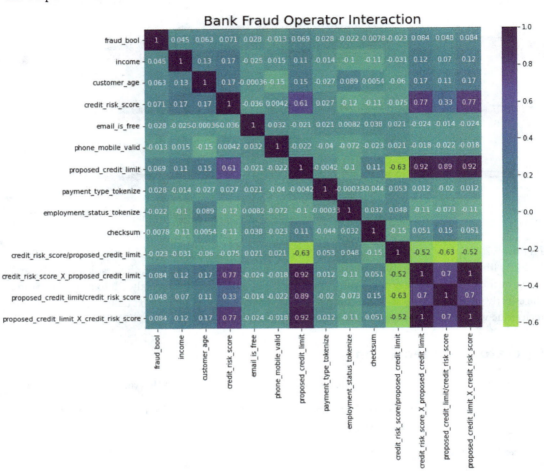

Figure 9.21 – Bank Fraud operator augmentation, correlogram plot

Figure 9.21 shows a strong relationship between three new vectors: `credit_risk_score_X_proposed_credit_limit` (multiply), `proposed_credit_limit/credit_risk_score` (divide), and `proposed_credit_limit_X_credit_risk_score` (multiply). Pluto implements the operator function correctly but still determines the benefit of the DL prediction accuracy.

The other five interaction tabular augmentation techniques are as follows:

- **Discretizing** is a method that uses **decision trees, equal width binding, equal frequency binding**, and **K-means clustering** to augment the tabular data. The discretization method depends on the AI model and the tabular data properties. Pluto recommends trying multiple approaches and evaluating their performance.

- The **quantile normalizing** method makes the distributions of the datasets comparable by transforming them so that they have the same cumulative distribution value.

- The **haversine distance** calculates the shortest distance between two angular points, such as the Earth's surface. Tabular augmentation also uses the **Euclidean, Mahalanobis**, and **Minkowski** distance algorithms.

- The **technical** indicator is one of the **specialty** methods in tabular augmentation. It uses technical analysis to help predict future price movements of securities or financial instruments. They are based on mathematical calculations of price, volume, and open interest.

- The **genetic** method, or **genetic** tabular augmentation, is a type of ML technique that uses evolutionary algorithms to optimize the AI model. The concept is to create a population of candidate solutions, or **chromosomes**, for a problem, then evolve that population over time by applying genetic operations such as crossover, mutation, and selection. The goal is to find the best solution to the problem through natural selection.

Fun challenge

Pluto challenges you to implement two more interaction augmentations in the Python Notebook.

The next tabular augmentation class is **mapping** augmentation. Pluto will describe the mapping functions but not implement them in the Python Notebook.

Mapping augmentation

The mapping method uses ML and data analysis to summarize and reduce the dimensionality of data for augmentation. It can be done via unsupervised or supervised learning. Some examples of mapping methods include **eigendecomposition** and PCA. PCA is a statistical procedure that transforms a set of correlated variables into uncorrelated variables called principal components.

In the DeltaPy library, there are seven mapping methods for tabular augmentation. Pluto has done a few implementations in the Python Notebook, but he will not explain the coding here. The Python wrapper function is repetitive and can easily be applied to any mapping method. The functions are as follows:

- **Eigendecomposition (ED)** is a form of **PCA** for tabular augmentation. In ED, the **eigenvectors** are the covariance matrix of the data, and the corresponding **eigenvalues** represent the amount of variance by each component. ED includes **linear discriminant analysis (LDA), singular value decomposition (SVD)**, and **Markov chains**.

- **Cross-decomposition** methods, including **canonical correlation analysis (CCA)**, are used to uncover linear relationships between two pieces of multivariate tabular data. Various applications, such as dimensionality reduction, feature extraction, and feature selection, use the cross-decomposition method. The goal is to find a linear combination between tabular data variables.

- **Kernel approximation** methods are used in ML algorithms such as SVMs to transform the tabular data into a higher dimensional space where a linear boundary can be found to separate the classes. The **additive Chi2 kernel** is a specific **kernel approximation** method that measures the independence between two sets of variables.

- **Autoencoders** are used in various domains, such as image compression, anomaly detection, and for generating new data for tabular augmentation. We use autoencoders in the pre-training step for supervised learning tasks to improve the subsequent models' performance.

- **Manifold learning** is a class of techniques for non-linear dimensionality reduction, aiming to preserve the tabular data's underlying non-linear structure. **Locally linear embedding (LLE)** is one method in which the idea is to approximate the local linear relationships between data points in the high-dimensional space. The goal is to find a lower-dimensional representation of high-dimensional data that still captures the essential patterns and relationships in the tabular data.

- **Clustering** is a popular unsupervised ML technique for grouping similar tabular data points into clusters. Clustering methods help in identifying patterns and structure in tabular data.

- **Neighbouring** is the nearest neighbor method for supervised ML algorithms for classification and regression problems. It also is used in tabular augmentation. You can extend the nearest neighbor method to a more sophisticated version called **k-nearest neighbor (k-NN)** classification.

The next classification of tabular augmentation we'll look at is **extraction** augmentation.

Extraction augmentation

The extraction method is a process in time series analysis where multiple constructed elements are used as input, and a singular value is extracted from each time series to create new augmented data. This method uses a package called **TSfresh** and includes default and custom features. The output of extraction methods differs from the output of transformation and interaction methods, as the latter outputs entirely new time series data. You can use this method when specific values need to be pulled from time series data.

The DeltaPy library contains 34 extraction methods. Writing the wrapper functions for extraction is similar to the wrapper transformation functions. The difficulty is how to discern the forecasting's effectiveness from tabular augmentation. Furthermore, these methods are components and not complete functions for tabular augmentation.

Pluto will not explain each function, but here is a list of the extraction functions in the DeltaPy library: Amplitude, Averages, Autocorrelation, Count, Crossings, Density, Differencing, Derivative, Distance, Distribution, Energy, Entropy, Exponent,

Fixed Points, Fluctuation, Fractals, Information, Linearity, Location, Model Coefficients, Non-linearity, Occurrence, Peaks, Percentile, Probability, Quantile, Range, Shape, Size, Spectral Analysis, Stochasticity, Streaks, Structural, and Volatility.

The extraction method is the last tabular augmentation category. Thus, it is time for a summary.

Summary

Tabular augmentation is a technique that can improve the accuracy of ML models by increasing the amount of data used. It adds columns or rows to a dataset generated by existing features or data from other sources. It increases the available input data, allowing the model to make more accurate predictions. Tabular augmentation adds new information not currently included in the dataset, increasing the model's utility. Tabular augmentation is beneficial when used with other ML techniques, such as DL, to improve the accuracy and performance of predictive models.

Pluto downloaded the real-world Bank Fraud and World Series datasets from the *Kaggle* website. He wrote most of the code in the Python Notebook for visualizing large datasets using various graphs, such as histograms, heatmaps, correlograms, and waffle and joy plots. He did this because understanding the datasets is essential before augmenting them. However, he didn't write a CNN or RNN model to verify the augmentation methods because building a CNN model is a complex process worthy of a separate book.

The DeltaPy open source library contains dozens of methods for tabular augmentation, but it is a beta version and can be unstable to load. Still, Pluto demonstrated a few tabular augmentation techniques, such as the robust scaler, standard scaler, capping, regression, and operator methods.

Throughout this chapter, there were *fun facts* and *fun challenges*. Pluto hopes you will take advantage of these and expand your experience beyond this book's scope.

This is the last chapter of the book. You and Pluto have covered augmentation techniques for image, text, audio, and tabular data. As AI and generative AI continue to expand and integrate into the fabric of our life, data will play an essential role. Data augmentation methods are the best practical option to extend your datasets without the high cost of gathering and purchasing additional data. Furthermore, generative AI transforms how we work and play, such as OpenAI's GPT3, GPT4, Google Bard, and Stability.ai's Stable Diffusion. What you discussed about AI in boardrooms or classrooms last month will be outdated, but the data augmentation concepts and techniques remain the same.

You and Pluto have learned to code the augmentation techniques using wrapper functions and download real-world datasets from the Kaggle website. As new, better, and faster augmentation libraries are available, you can add to your collection or switch the libraries under the hood. What you implement may change slightly, but what you have learned about data augmentation will remain true.

I hope you enjoyed reading the book and hacking the Python Notebook as much as I enjoyed writing it. Thank you.

Index

www.packtpub.com

Subscribe to our online digital library for full access to over 7,000 books and videos, as well as industry leading tools to help you plan your personal development and advance your career. For more information, please visit our website.

Why subscribe?

- Spend less time learning and more time coding with practical eBooks and Videos from over 4,000 industry professionals

- Improve your learning with Skill Plans built especially for you

- Get a free eBook or video every month

- Fully searchable for easy access to vital information

- Copy and paste, print, and bookmark content

Did you know that Packt offers eBook versions of every book published, with PDF and ePub files available? You can upgrade to the eBook version at www.packtpub.com and as a print book customer, you are entitled to a discount on the eBook copy. Get in touch with us at customercare@ packtpub.com for more details.

At www.packtpub.com, you can also read a collection of free technical articles, sign up for a range of free newsletters, and receive exclusive discounts and offers on Packt books and eBooks.

Other Books You May Enjoy

If you enjoyed this book, you may be interested in these other books by Packt:

Hands-On Graph Neural Networks Using Python

Maxime Labonne

ISBN: 978-1-80461-752-6

- Understand the fundamental concepts of graph neural networks
- Implement graph neural networks using Python and PyTorch Geometric
- Classify nodes, graphs, and edges using millions of samples
- Predict and generate realistic graph topologies
- Combine heterogeneous sources to improve performance
- Forecast future events using topological information
- Apply graph neural networks to solve real-world problems

The Ultimate Guide to ChatGPT and OpenAI

Valentina Alto

ISBN: 978-1-80512-333-0

- Understanding of generative AI concepts from basic to intermediate level
- Focus on GPT architecture for generative AI models
- Maximize ChatGPT value with an effective prompt design
- Explore applications and use cases of ChatGPT
- Use OpenAI models and features via API calls
- Build and deploy generative AI systems with Python
- Leverage Azure infrastructure for enterprise-level use cases
- Ensure Responsible AI and ethics in generative AI systems

Packt is searching for authors like you

If you're interested in becoming an author for Packt, please visit `authors.packtpub.com` and apply today. We have worked with thousands of developers and tech professionals, just like you, to help them share their insight with the global tech community. You can make a general application, apply for a specific hot topic that we are recruiting an author for, or submit your own idea.

Share your thoughts

Now you've finished *Data Augmentation with Python*, we'd love to hear your thoughts! Scan the QR code below to go straight to the Amazon review page for this book and share your feedback or leave a review on the site that you purchased it from.

`https://packt.link/r/1-803-24645-6`

Your review is important to us and the tech community and will help us make sure we're delivering excellent quality content.

Download a free PDF copy of this book

Thanks for purchasing this book!

Do you like to read on the go but are unable to carry your print books everywhere?

Is your eBook purchase not compatible with the device of your choice?

Don't worry, now with every Packt book you get a DRM-free PDF version of that book at no cost.

Read anywhere, any place, on any device. Search, copy, and paste code from your favorite technical books directly into your application.

The perks don't stop there, you can get exclusive access to discounts, newsletters, and great free content in your inbox daily

Follow these simple steps to get the benefits:

1. Scan the QR code or visit the link below

https://packt.link/free-ebook/9781803246451

1. Submit your proof of purchase
2. That's it! We'll send your free PDF and other benefits to your email directly

www.ingramcontent.com/pod-product-compliance
Lightning Source LLC
Chambersburg PA
CBHW060922060326
40690CB00041B/2962